Focus on Genetic Screening Research

FOCUS ON GENETIC SCREENING RESEARCH

SANDRA R. PUPECKI
EDITOR

Nova Science Publishers, Inc.
New York

For permission to use material from this book please contact us:
Telephone 631-231-7269; Fax 631-231-8175
Web Site: http://www.novapublishers.com

Library of Congress Cataloging-in-Publication Data
Available upon request.

ISBN 1-60021-229-8

Published by Nova Science Publishers, Inc. ✣ New York

Contents

Preface

Gene tests (also called DNA-based tests), the newest and most sophisticated of the techniques used to test for genetic disorders, involve direct examination of the DNA molecule itself. Other genetic tests include biochemical tests for such gene products as enzymes and other proteins and for microscopic examination of stained or fluorescent chromosomes. Genetic tests are used for several reasons, including: Carrier screening, which involves identifying unaffected individuals who carry one copy of a gene for a disease that requires two copies for the disease to be expressed; Preimplantation genetic diagnosis prenatal diagnostic testing newborn screening; Presymptomatic testing for predicting adult-onset disorders such as Huntington's disease; Presymptomatic testing for estimating the risk of developing adult-onset cancers and Alzheimer's disease; Confirmational diagnosis of a symptomatic individual forensic/identity testing. In gene tests, scientists scan a patient's DNA sample for mutated sequences. A DNA sample can be obtained from any tissue, including blood. For some types of gene tests, researchers design short pieces of DNA called probes, whose sequences are complementary to the mutated sequences. These probes will seek their complement among the three billion base pairs of an individual's genome. If the mutated sequence is present in the patient's genome, the probe will bind to it and flag the mutation. Another type of DNA testing involves comparing the sequence of DNA bases in a patient's gene to a normal version of the gene. This new book gathers important new research in this field.

In chapter I, Cystic fibrosis (CF) mutation analysis on 500 CF patients, characterized 83 different mutations (20 so far specific to our population) accounting for 92% of CF genes and generating 111 different genotypes. Seven mutations were common [p.F508del (54.5%), c.621+1G>T (6.3%), p.G542X (3.9%), p.N1303K (2.3%), c.2789+5G>A (2.0%), c.2183AA>G (1.5%), p.E822X (1.8%)], 16 showed frequencies between 0.4% and 1.1%, while the remaining (60) were very rare (1 to 3 alleles). Denaturing gradient gel electrophoresis (DGGE) screening of 15 exons (3, 4, 7, 9, 10, 11, 12, 13, 14b, 16, 17b, 19, 20, 21, and 23) detected 86% of CF alleles. Haplotypes for eight diallelic and three microsatellite markers have been characterized for the common, a few rare and novel Greek mutations. In order to test the contribution of the CFTR gene in male infertility, we carried out mutation testing on subjects suffering from the monosymptomatic form of CFTR. Mutation testing was carried out on 134 male infertility subjects including 88 suffering from azoospermia (Congenital bilateral absence of vas deferens-CBAVD and Obstructive azoospermia-OA) and

46 with severe oligospermia. The results showed that 56.2% of CBAVD males carried one mutation/variant in the CFTR gene while 27.1% were compound heterozygotes for one mild (belonging to class IV or V) and one severe mutation. Of the OA males, 27.5% had at least 1 CF allele while only 13% of the severe oligospermia males were carriers. Carrier screening was initiated, with emphasis on individuals and couples in high-risk groups –with a family history of CF, one partner with CF, and couples with male infertility seeking in vitro fertilization (IVF). Mutation analysis on 900 individuals (175 couples, 91 unaffected CF siblings, 285 CF family relatives and 174 general population subjects), identified a total of 188 heterozygotes the majority amongst siblings and relatives of index cases. Of the 175 couples, 23 had advanced ongoing pregnancies upon request for CF screening. In total six couples were found requiring prenatal diagnosis. Results of 167 prenatal diagnoses (including 110 due to bowel echogenicity), testing a total of 57 different parental genotypes, are reported. One hundred and sixty-seven prenatal tests resulted in 41 affected, 80 heterozygous, 46 normal fetuses and one incomplete diagnosis. Of the echogenic bowel fetuses, 15 had one detectable mutation and in 4 cases homozygosity with severe mutations was detected.

Prenatal genetic screening tests are used to analyze the number of chromosomes and to detect the chromosomal abnormalities such as Down syndrome during the early period of pregnancy. Genetic tests enable couples to be informed beforehand about the health of their child and offer the opportunity to terminate an affected pregnancy. However, genetic screening testing following an amniocentesis procedure in the prenatal period, and a positive test result cause anxiety in pregnant women. In chapter II, this study is carried out as descriptive to evaluate the distress of pregnant women in relation to genetic testing, their anxiety level and the effects of risk factors on anxiety. In this study 130 pregnant women who had undergone genetic testing following an amniocentesis were interviewed. A questionnaire was used for data collection. Anxiety of the women was assessed using the Spielberger's state-trait anxiety inventory (STAI). Completing the questionnaire form and anxiety scales lasted about half an hour. The majority of the women who underwent amniocentesis expressed distress prior to genetic testing. The mean trait anxiety scores of women before receiving the test results was 43.72 and the state anxiety scores of women after the test result was 36.05. The association of state anxiety level of pregnant women after they received the test results indicated, those women who had no living child, who had a child with abnormality, who reported one or more spontaneously abortion history, who were younger, and who have a positive test result revealed significantly higher anxiety level. In contrast, there was no significant association between the trait anxiety of women and the education level, having a consanguine marriage, and the need for counseling following the test results. In this study it was identified that women referred for genetic testing after an amniocentesis procedure experienced distress before testing, they were anxious and psychologically affected by undergoing amniocentesis and genetic testing, even if the test results were negative. Perceiving not to be informed in detail about genetic testing and amniocentesis raised their anxiety level. The results revealed the importance of a comprehensive genetic counseling for women who underwent amniocentesis and genetic testing.

The Jewish population has remained isolated for centuries, and as a direct result, many genetic disorders are relatively frequent among Jews. For most of the severe autosomal recessive diseases no treatment is available and prevention is therefore the only means to reduce the morbidity. The Tay - Sachs disease carrier screening prevention program was initiated in 1971 among Jews of Ashkenazi origin and remained up to now very successful. In the last decade, the possibility of carrier detection has also become available for most of the other relatively frequent disorders existing in the different Jewish communities, leading to the possibility of expanding genetic screening programs. According to the guidelines for population screening first delineated by Wilson and Jungner for the World Health Organization, the three major criteria for genetic screening for reproductive purposes are (1) the disease is severe and occurs predominantly in a well-defined population, (2) the detection of heterozygotes is possible using simple, accurate and inexpensive technology and (3) it is possible to prevent the birth of affected children. While the application of these criteria seems easy, the definitions of severity and frequency may be controversial. For instance, the tests are often offered to couples of mixed origin even though the risk is very low. In addition, the genetic screening offered to the Jewish population includes disorders such as Gaucher disease or deafness, which for many are not considered severe enough to justify population screening. As described in chapter III, a screening program aimed at the Jewish population should be based upon criteria accepted by the communities and must be preceded by genetic counseling, emphasizing its aims and limitations.

Human blood platelets represent cells with a variety of important functions in hemostatic, immunological and inflammatory processes. Glycoproteins (GP) and other receptors are expressed on the cell surface and are involved in platelet adhesion to matrix proteins or cells. Chapter IV is intended to give a summary of the current knowledge about the clinical relevance of single nucleotide polymorphisms (SNPs) and mutations in platelet receptor genes. Some SNPs represent the genetic basis of the human platelet alloantigen (HPA) systems. The gene variants are considered in the context of platelet disfunction as well as in relation to the pathogenesis of vascular diseases and thromboembolic events. Other polymorphic adhesion receptors expressed on platelets are, for example, P-selectin, PSGL-1 and PECAM-1. Results from case/control studies including patients with vascular diseases or thromboembolic events are summarized. The review also considers the genetic typing techniques currently available for the platelet receptor polymorphisms. A variety of techniques for genotyping platelet receptor polymorphisms, mostly HPA SNPs, have been developed and some of them are nowadays routinely used in specialized clinical diagnosis. Prenatal HPA-typing mostly done on fetal DNA obtained by amniocentesis or umbilical cord vein puncture represents a relevant aspect in the early diagnosis of neonatal alloimmune thrombocytopenia (NAIT). The use of fetal DNA circulating in the maternal plasma is a challenge for future developments. The HPA-typing approaches are predominantly performed as individual testing and are only practicable for sample numbers in a low throughput range. Other techniques are currently developed on the basis of microarray platforms and will allow a complex typing of large numbers (>100) of reasonable SNPs in a single analysis. In future, these techniques may be used for both, the low throughput individual testing in clinical diagnosis and the high throughput screening of blood donors for certain constellations of blood cell antigens.

Rapid advances in DNA microarray analysis technology provide new tools for establishing prognostic predictors of various diseases. In such tools, it is desired to find significant genes selectively for constructing the prognostic model and also is necessary to eliminate nonspecific genes or genes with error for construction of the model. There are two kinds of methods, filter and wrapper approaches, for gene selection. In a previous study, the authors developed the projective adaptive resonance theory (PART) as filtering approach and boosted fuzzy classifier with SWEEP operator (BFCS) as a wrapper approach. In chapter V, they applied PART to gene expression data in order to eliminate nonspecific genes and genes with error. Next, the genes selected by PART were subjected to our fuzzy neural network (FNN) combined with SWEEP operator method (FNN-SWEEP) for construction of a cancer class prediction models. The results in predictor model were evaluated through comparison with a conventional screening method (signal-to-noise (S2N) method) or nearest shrunken centroids (NSC) method. The FNN-SWEEP predictor with PART screening could discriminate classes of acute leukemia in blinded data with 97.1% accuracy and classes of lung cancer with 90.0% accuracy, while the predictor with S2N was only 85.3% and 70.0% or the predictor with NSC was 88.2% and 90.0%, respectively. Next, BFCS was applied to microarray data of brain tumor (central nervous system tumor). The results in the predictor model were evaluated through comparison with a conventional screening method, such as support vector machine (SVM), FNN-SWEEP, k-nearest neighbor (kNN), multiple regression analysis (MRA), and weighted voting (WV). The accuracy of the BFCS method reached 71.3% in the 4-input models. On the other hand, SVM, FNN-SWEEP, kNN, MRA, and WV showed an accuracy of 54.3% to 66.3%, which was lower than that of BFCS. The results have proven that PART and BFCS were superior to other conventional methods, respectively.

Microstellites are DNA sequences consisting of short repeats of highly variable number. Hundreds thousands of such sequences have been found throughout the human genome. Microsatellites are extremely powerful as markers in distinguishing alleles at linked loci and thus is widely used for linkage studies, segregation analysis and determining allele loss or gain in tumors. In chapter VI, several extended applications of microsatellites are markers will be introduced:

1. Using multiple highly informative microsatellites for screening of large (megabase) deletions.
2. Using information obtained from loss of heterozygosity analysis for
   a. tracing parental origin of alleles bearing *de novo* mutations
   b. determining mutation-carrying alleles for presymptomatic / prenatal diagnosis for at-risk individuals
   c. finding cellular origin of tumor cells
   d. verification of the identity of cells cultured from tumors.

In: Focus on Genetic Screening Research
Editor: Sandra R. Pupecki, pp. 1-26
ISBN 1-60021-229-8

Chapter I

Screening for Cystic Fibrosis Mutations in Southern European Countries: Methods for Molecular Diagnosis, Prenatal Diagnosis and Carrier Identification Amongst High-Risk Individuals- the Greek Experience

Emmanuel Kanavakis [*] ***and Maria Tzetis*** *
University of Athens, St. Sophia's Children's Hospital,
Athens, 11527, Greece

Abstract

Cystic fibrosis (CF) mutation analysis on 500 CF patients, characterized 83 different mutations (20 so far specific to our population) accounting for 92% of CF genes and generating 111 different genotypes. Seven mutations were common [p.F508del (54.5%), c.621+1G>T (6.3%), p.G542X (3.9%), p.N1303K (2.3%), c.2789+5G>A (2.0%), c.2183AA>G (1.5%), p.E822X (1.8%)], 16 showed frequencies between 0.4% and 1.1%, while the remaining (60) were very rare (1 to 3 alleles). Denaturing gradient gel electrophoresis (DGGE) screening of 15 exons (3, 4, 7, 9, 10, 11, 12, 13, 14b, 16, 17b, 19, 20, 21, and 23) detected 86% of CF alleles. Haplotypes for eight diallelic and three microsatellite markers have been characterized for the common, a few rare and novel Greek mutations. In order to test the contribution of the CFTR gene in male infertility, we carried out mutation testing on subjects suffering from the monosymptomatic form of CFTR. Mutation testing was carried out on 134 male infertility subjects including 88 suffering from azoospermia (Congenital bilateral absence of vas deferens-CBAVD and

* Dept of Medical Genetics, University of Athens, St. Sophia's Children's Hospital, Athens, 11527, Greece; Fax: +30210 7795553 ; Email: ekanavak@cc.uoa.gr; mtzeti@cc.uoa.gr

Obstructive azoospermia-OA) and 46 with severe oligospermia. The results showed that 56.2% of CBAVD males carried one mutation/variant in the CFTR gene while 27.1% were compound heterozygotes for one mild (belonging to class IV or V) and one severe mutation. Of the OA males, 27.5% had at least 1 CF allele while only 13% of the severe oligospermia males were carriers. Carrier screening was initiated, with emphasis on individuals and couples in high-risk groups –with a family history of CF, one partner with CF, and couples with male infertility seeking in vitro fertilization (IVF). Mutation analysis on 900 individuals (175 couples, 91 unaffected CF siblings, 285 CF family relatives and 174 general population subjects), identified a total of 188 heterozygotes the majority amongst siblings and relatives of index cases. Of the 175 couples, 23 had advanced ongoing pregnancies upon request for CF screening. In total six couples were found requiring prenatal diagnosis.

Results of 167 prenatal diagnoses (including 110 due to bowel echogenicity), testing a total of 57 different parental genotypes, are reported. One hundred and sixty-seven prenatal tests resulted in 41 affected, 80 heterozygous, 46 normal fetuses and one incomplete diagnosis. Of the echogenic bowel fetuses, 15 had one detectable mutation and in 4 cases homozygosity with severe mutations was detected.

Key words: Cystic Fibrosis/ mutation screening/ CBAVD/ obstructive azoospermia/ oligospermia/ ICP/ prenatal diagnosis/ echogenic bowel

Introduction

Cystic fibrosis (CF; OMIM 219700, CFTR; OMIM 602421) is a severe autosomal recessive genetic disease, affecting 1 in 2500 Caucasians with 4% being asymptomatic carriers of mutations in the gene [Welsh et al., 2001; Kerem et al., 1989; Riordan et al., 1989; Rommens et al., 1989]. The median age of diagnosis is 7 months with more than 60% of the affected individuals diagnosed by 1 year of age. Without treatment, CF is typically a fatal disorder in infancy and childhood. The concentration of salt in the sweat of CF patients is usually elevated (>60mEq/l) and it is used as a diagnostic test. With currently improved treatment the median age of survival has increased to more than 30 years. However patients continue to have a considerable burden of illness and die prematurely due to chronic pulmonary damage from excessive inflammation and chronic infection [Welsh et al., 2001]. There is substantial variability in the clinical expression of the disease, with patients symptomatic at birth, while others not presenting symptoms for months or even years. More than 96% of men with CF have anomalies in the Wolffian structures causing obstructive azoospermia and infertility with the body and tail of the epididymis and seminal vesicles atrophic or absent. In approximately 85% of CF patients, the exocrine pancreas is affected (PI, pancreatic insufficiency). In the remaining 15% of patients, the exocrine pancreas is functionally adequate (PS, pancreatic sufficiency). The gene which was characterized in 1989 [Kerem et al., 1989; Riordan et al., 1989; Rommens et al., 1989], is located on chromosome 7q31.2, encoding a glycosylated transmembrane protein. The 1480 amino acid transmembrane protein has a symmetrical structure of two repeats composed of a transmembrane region (TMD) containing six transmembrane helices (TM) and a nucleotide binding domain (NBD), separated by a large hydrophilic regulatory (R) domain. These features are characteristic of a large family of ATP-Binding Cassette Transporters (ABC transporters) found in prokaryotes and eukaryotes. The difference is that the R domain is unique to the CFTR protein [Vankeerberghen et al., 2002]. The CFTR protein is a chloride ion channel, regulating salt and water transport and is expressed in the apical membrane of epithelial exocrine cells. Besides its function as a chloride channel, CFTR regulates other apical membrane conductance pathways. Activity of the epithelial sodium channel is inversely related to CFTR activity. CFTR is thought to regulate the outwardly rectifying chloride channel (ORCC) by facilitating cellular release of ATP. CFTR might also transport HCO_3 or regulate its transport through epithelial cell membranes. CFTR contributes to normal pancreatic secretion by diluting and alkalinizing the pancreatic juice to prevent obstruction of the smaller pancreatic ducts with proteinaceous plugs, therefore reduced CFTR function might also contribute to ductal plugging in pancreatic disease other than CFTR [Cohn 2005]. Since the initial characterization of the predominant mutation in the Caucasian population, F508del [Kerem et al., 1989], more than 1000 mutations have been reported to the Cystic Fibrosis Genetic Analysis Consortium [CFGAC: http://www.genet.sickkids.on.ca/cftr/], most of which are presumed to be pathogenic. These mutations express their effects through a variety of molecular mechanisms that include either the failure of production of CFTR protein or the production of a mutant protein with no or variable residual function and are thus subdivided into six classes: [Zielenski, 2000; Vankeerberghen et al., 2002]. Class I includes nonsense, frameshift and splice site mutations, giving rise to the formation of premature termination signals. Such proteins are often unstable and are expected to be degraded resulting in minimal or no functional CFTR chloride activity in the affected epithelia. Class II contains missense and amino acid deletion mutations including F508del, and is associated with improper maturation of the corresponding CFTR proteins. These immature mutant proteins will remain associated with molecular chaperones and will ultimately be degraded and as a result will not appear at the cell surface. Class III are missense mutations, mostly located in the nucleotide binding domains. They affect the regulation of the CFTR channel by interfering with the phosphorylation of the R domain and binding and hydrolysis of ATP at both nucleotide binding domains. They result in a decrease of the net chloride transport activity of the channel. Class IV mutations are also missense located in the pore of the channel and give rise to a CFTR channel with defective conductive properties. Class V are missense and alternate splicing mutations mostly causing reduced CFTR synthesis. Especially with splice-site mutations, normal splicing can still occur in a small proportion of these transcripts and a small amount of functional CFTR protein will be produced resulting in a mild phenotype. Class VI are mutations that affect CFTR protein regulation of other channels or decrease the apical stability of the protein. Class I-III and VI mutations are characterized as severe and are associated with pancreatic insufficiency since no functional CFTR is present at the apical membrane, whereas patients with rarer class IV-V mutations with some residual function are mostly pancreatic sufficient and characterized as mild. Although classic CF is diagnosed in infancy or early childhood, atypical cases of CF may be diagnosed in adults with mild monosymptomatic diseases such as congenital bilateral absence of the vas deferens (CBAVD, CAVD), idiopathic chronic or acute and recurrent pancreatitis (ICP and ARP), sinusitis, diffuse bronchiectasis, chronic obstructive pulmonary disease or asthma [Kanavakis et al., 1998; Tzetis et al., 2001a; Noone and Knowles, 2001; Casals et al., 2004; Claustres, 2005; Ahmed et al., 2003]. Individuals with atypical CF phenotypes have normal or borderline sweat chloride tests and are often compound heterozygotes carrying one severe and one mild mutation/variant (class IV and V mutations), or are homozygous for the IVS8-5T variant. Near the 3' region of intron 8 in the splice acceptor site, there is a tract of 5, 7, or 9 thymidine residues. Usually 7 or 9 thymidines are present, which allows normal gene splicing and CFTR production. When 5 thymidine residues are present-the IVS8-5T variant-abnormal splicing occurs, which leads to the skipping of exon 9 and the presence of approximately only 5-10% of the normal CFTR messenger RNA. The CFTR mRNA deleted for exon 9 maintains an open reading frame and would encode a CFTR isoform that is missing amino acids 404-464 of the first nucleotide binding domain (NBD1). This protein is not properly processed and is not capable of generating Cl^- conductance in response to cAMP [Strong et al., 1993]. The 5T variant is present on about 5% of the alleles in the general population and is associated with male infertility and/or mild pulmonary symptoms. This allele shows incomplete penetrance which limits its prognostic or diagnostic value for individual patients [Groman et al., 2004].

CF is an autosomal recessive disorder and affected individuals typically inherit one mutated allele from each parent. Therefore couples who have had one affected offspring are likely to be CF carriers with a 25% recurrence risk in each future pregnancy. Healthy siblings of an affected person have a two-thirds chance of being CF carriers. Second- and third-degree relatives of affected individuals have a 50% and 25% risk of being carriers, respectively. Complex alleles, in which more than one mutation is present in cis, have been observed [Zielenski and Tsui 1995]. F508del shows a higher frequency in northern European in comparison to southern European populations [Welsh et al., 2001], where the percentage of non-F508del chromosomes associated with CF is high. The frequencies and types of mutations found in populations vary according to the geographic and ethnic origin of the population studied. In populations with high heterogeneity, mutation analysis can be facilitated by association between haplotypes for diallelic and microsatellite polymorphisms for each CF mutation. Several highly polymorphic microsatellite and diallelic markers have been described within the CFTR gene and have been used in association analysis with CF mutations in other populations [Dork et al., 1992; Claustres et al., 1996]. Accurate knowledge of CF mutations in a specific population provides information for CF prevention programs applicable through heterozygote screening and prenatal diagnosis.

The aim of genetic screening for CF is to aid the process of reproductive decision making. This is principally achieved by identifying carrier couples who can have prenatal diagnosis and selective termination of pregnancy. Because it is not feasible to determine the carrier status of all persons of reproductive age in the population, selective screening by focusing on the relatives of the index case can be used as an alternative approach. This has the advantage of detecting a higher ratio of carriers and restriction of assays to the mutant alleles in the index case at least for first degree relatives. CF carrier testing should also be offered to males with a family history of unilateral or bilateral absence of vas deferens and their partners. Prenatal diagnosis for CF is additionally offered to pregnancies diagnosed with echogenicity of the fetal bowel. Fetal echogenic bowel, with a prevalence of 0.6-1.8% in second trimester pregnancies, is defined as echogenicity of the fetal bowel approaching that of the surrounding bone. It is described as a normal variant but also in association with chromosomal abnormalities (~9%), cystic fibrosis (~2%), intrauterine growth restriction (~8%), intestinal obstruction, congenital infection (~3%), and intra-amniotic bleeding [Monaghan et al., 1999]. In pregnancies with a 25% risk of cystic fibrosis the echogenic bowel may represent meconium impaction and solidification in the small bowel. Procedure for CF screening in these pregnancies is to test for 85-90% of CF alleles found in the specific population. Screening is done within limited time constraints due to advanced age of pregnancy.

This study aims to offer a strategy for CFTR mutation screening for populations with high molecular heterogeneity using the Greek population as one such example. We report results of the molecular analysis of 500 Greek classic CF patients, including haplotype information, and CF mutation screening for atypical CF on 134 male subjects suffering from infertility (CBAVD, OA and Oligospermia). A total of 152 prenatal diagnoses for pregnancies at a 1 in 4 risk, 15 for couples with male infertility problems and 110 due to fetal bowel echogenicity are reported. The outcome of carrier detection offered to relatives of CF patients and their partners, couples with male infertility seeking IVF and members of the general population (total of 900 individuals) is also discussed.

Materials and Methods

Molecular Characterization and Haplotype Analysis

A total of 500 unrelated CF patients of Greek origin (1000 chromosomes), the majority of who attend the Cystic Fibrosis Unit of the St. Sophia's Children's Hospital and children and adults from other Cystic Fibrosis units in Greece, were included in this study. Diagnostic criteria involved positive sweat tests and typical clinical findings of pulmonary and gastrointestinal disease. The 27 exons and neighboring intronic regions of the CFTR gene were assessed by denaturing gradient gel electrophoresis (DGGE). Haplotype analysis for the diallelic sites –XV2c-KM19-D9-J44-M470V-T854-TUB20– was conducted on 261 p.F508del, 116 non-p.F508del and 51 normal chromosomes. For the three microsatellite loci, IVS8CA-IVS17BTA-IVS17BCA, analysis was performed on 146 p.F508del, 85 non-p.F508del and 136 normal alleles.

Monosymptomatic CFTR Screening

Male Infertility

Congenital Bilateral Absence of Vas Deferens (CBAVD) - Obstructive Azoospermia (OA) – Oligospermia

Male subjects (n=134) suffering from infertility with azoospermia (48 CBAVD and 40 OA) or severe oligospermia (46) were studied. The patients were referred to us from various infertility clinics. The diagnosis of CBAVD was based on the small volume of ejaculate and complete absence of spermatozoa, and the typical biochemical changes found in vas deferens agenesis [Kanavakis et al., 1998]. The presence of the vas deferens in the 40 males with OA was confirmed by clinical examination. Renal ultrasound was not performed on the CBAVD patients in order to exclude renal agenesis as the cause of CBAVD. Analysis included DGGE screening of 15 exons, and sequencing if the mutation(s) did not present with a known pattern.

None of the monosymptomatic disease patients had any of the classic symptoms of CF including serious pulmonary involvement.

Heterozygote Screening

Carrier identification through DGGE screening of 15 exons (for 86% of detectable CF alleles) was completed for 900 individuals. These included 175 couples (350 individuals) belonging to a high-risk group with a family history of CF or seeking in vitro Fertilization (IVF) due to male infertility, 91 unaffected CF siblings, 285 CF family relatives (not married therefore not part of couple screening) and 174 general population subjects.

Prenatal Diagnosis

Prenatal diagnosis (PND) was offered to 150 families (152 diagnoses), the majority of which had at least 1 affected child and to seven couples (15 prenatal diagnoses), with male infertility problems. Chorionic villus (CVS) biopsy was the study material in 139 cases and amniotic fluid (AF) in 28 cases. Six cases involved twin pregnancies, four from IVF couples and one triplet. Pre-implantation Genetic Diagnosis (PGD) [Vrettou et al., 2002] was performed in two IVF couples followed by 1 triplet pregnancy and one singleton. The PGD results were confirmed by PND. In two cases we observed parental alleles carrying two mutations in cis, and four cases involving compound heterozygous parents, two asymptomatic and two CBAVD.

A combination of DGGE-CFTR mutation screening and polymerase chain reaction–mediated site-directed mutagenesis (PMS) [Friedman et al., 1991; Tzetis et al., 1997] for the confirmation of the mutation(s) was used for each diagnosis. Each prenatal diagnosis included re-testing of sample DNA from parents and healthy and/or affected child. The three microsatellite markers (IVS8CA, IVS17BTA and IVS17BCA) used for indirect analysis of CF mutation(s), confirmation of paternity and possible maternal contamination of the fetal DNA sample were amplified for all the family members.

Prenatal Diagnosis Due to Fetal Bowel Hyperechogenicity

A total of 74 amniotic fluid samples (including parental DNA in 37 cases) and 36 couples were tested for CF mutations. None of the couples had a prior medical history of CF.

The cases were referred from different maternity hospitals and gynecologists (performing the ultrasound) to the Department of Medical Genetics of the University of Athens between 1999 -2005. The grade of echogenicity and criteria for sonographic diagnosis used by these centers was heterogeneous and often did not accompany the sample to be tested for CF. When amniocentesis for prenatal karyotyping to exclude aneuploidy was performed- CF testing included parental and fetal DNA to avoid false negative results due to non paternity. If the parents wanted to avoid an invasive procedure, parental carrier status was initially determined and amniocentesis offered to couples with both members CF carriers.

The average gestational age at amniocentesis was 20.5 weeks (range 20–23 weeks) for the 74 cases of fetal bowel hyperechogenicity and amniotic fluid was the study material. DNA from both or one of the parents when possible was incorporated in the analysis which included DGGE screening of 15 exons, confirmation with PMS or sequencing if the mutation(s) did not present with a known pattern. The three microsatellite markers (IVS8CA, IVS17BTA and IVS17BCA) were used for confirmation of paternity and possible maternal contamination of the fetal DNA sample.

The study was approved by the hospital's review board and all participants provided informed consent.

DNA Mutation Analysis

Genomic DNA was extracted from white blood cells, CVS and AF samples by standard isolation procedures using QIAamp DNA blood mini kit (Qiagen, Hilden, Germany) [Kanavakis et al., 2003]. The presence of mutations in the 27 exons of the CFTR gene and neighbouring intronic regions of the CFTR gene was assessed by denaturing gradient gel electrophoresis (DGGE) after simple or multiplex PCR of patient DNA [Fanen et al., 1992]. The intron 8 poly (T) variants of the CFTR gene were analyzed as described [Friedman et al., 1997]. All DNA samples showing a shift in DGGE mobility and not presenting a pattern of a known mutation were re-amplified from genomic DNA. Direct sequencing was performed using the VISGEN Long Read Tower system-V3.1 and the Cy5/Cy5.5 Dye primer bi-directional sequencing kit according to the manufacturer's instructions (Visible Genetics Inc., Toronto, Canada). All sequencing primers for the generation of the PCR products additionally contained universal 5' M13 sequences [Forward: 5'-GTAAAACGACGGCCAGT-3'; Reverse: 5'-CAGGAAACAGCTATGAC-3']. PCR primers for the CFTR gene exons were as previously described [Zielenski et al., 1991]. Common mutations in prenatal diagnoses were confirmed with the PMS methodology [Friedman et al., 1991; Tzetis et al., 1997].

Haplotype Analysis

For the diallelic sites, the PCR conditions and primers used are as described [Dork et al., 1992]. For haplotype analysis, restriction digests of the RFLP-PCR products were separated by agarose gel electrophoresis. Allele 1 represents the absence of a restriction site while allele 2 its presence. The GATT polymorphism was detected by the difference in mobility of the PCR product on 10% polyacrylamide gels. Allele 1 denotes 7 repeats, allele 2 six and allele 3 eight repeats. Microsatellites IVS8CA, IVS17BTA and IVS17BCA were analysed in a multiplex fluorescent PCR using the conditions reported previously [Morral et al., 1992]. All forward primers were fluorescently Cy5.5 labeled (Proligo LLC, Boulder, CO, USA). An aliquot of the reaction was mixed with a loading dye and 100 bp and 350 bp size markers (VISGEN Long Read Tower system-V3.1, Visible Genetics, Inc., Toronto, Canada), heated at 95°C for 5 min and cooled on ice. It was subsequently separated on a 6% denaturing polyacrylamide gel. Allele fragment sizes were determined in comparison with external size markers by an automated DNA sequencer and analyzed using the Fragment Analysis Software (Visible Genetics, Inc., Toronto, Canada).

Haplotypes were assigned where phase could be established, by typing of both parents and affected child in each CF pedigree.

Results

Molecular Characterization

In the sample of 500 Greek CF patients 83 different mutations were detected including three complex alleles, and 20 mutations so far only identified in the Greek population (table 1). These 83 mutations account for 92% of CF genes generating 111 different genotypes, including 10 different homozygote genotypes with the following mutations: p.F508del, c.621+1G>T, p.G542X, p.R1158X, p.I507del, p.W496X, p.E822X, c.2789+5G>A, p.G85E and p.R1070Q. Seven mutations were quite common belonging (except for p.E822X) to a group that are common worldwide with p.F508del accounting for 54.5%. Sixteen show frequencies between 0.4% and 1.1% while the remaining 60 were very rare found in 1–3 chromosomes.

Table 1. Frequencies of CFTR mutations identified in 500 Greek patients with cystic fibrosis, diagnosed on a clinical basis

Mutation	Location Exon/intron	No. of chromosomes (Frequency %)	Microsatellite haplotype[b]	Ethnic prevalence[c]
p.F508del	10	545 (54.5)	25-31/32-13	Pan-European
c.621+1G>T	intron 4	63 (6.3)	23-32-13	Welsh
p.G542X	11	39 (3.9)	25-33/34-13	Mediterranean
p.N1303K	21	23 (2.3)	25/26-32-13	Mediterranean
c.2789+5G>A	intron 14b	20 (2.0)	8-18/28-18	Mediterranean
p.E822X[a]	13	18 (1.8)	17-30-13	Greek
c.21833AA>G	13	15 (1.5)	17-29/31-13	Hispanic
p.R1158X	19	11 (1.1)	25-34-13	Mediterranean
p.G85E	3	9 (0.9)	18-8/24-18/13	
p.R334W	7	8 (0.8)		Multi-ethnic
c.574delA	4	8 (0.8)		
p.W1282X	20	8 (0.8)	18-8-17	Ashkenazi Jewish
c.1677delTA	10	7 (0.7)		Russian
c.3120+1G>A	intron 16	7 (0.7)		African-American/ Arabian
p.W496X	10	7 (0.7)	18-32-13	
c.3272-26A>G	intron 17a	7 (0.7)	18-8-18	Mediterranean
p.I507del	10	6 (0.6)	26-32-13	Hispanic
c.711+3A>G	intron 5	6 (0.6)		Mediterranean
c.621+3A>G[b]	intron 4	6 (0.6)		Mediterranean
p.R1070Q	17b	5 (0.5)		Bulgarian
p.R1070Q+S466X	17b and 10	4 (0.4)		Bulgarian

Table 1. Continued

Mutation	Location Exon/intron	No. of chromosomes (Frequency %)	Microsatellite haplotype[b]	Ethnic prevalence[c]
c.1525-1G>A	intron 9	4 (0.4)		Mediterranean
p.L137P[a]	4	4 (0.4)		Greek
p.A46D[a]	2	3 (0.3)	25-33-13	Greek
c.3120+1kbdel8.5kb	intron 16	3 (0.3)		Arabian
c.2184insA	13	3 (0.3)		
p.G551D	11	3 (0.3)		Celtic
p.D110H	4	3 (0.3)		Mediterranean
p.C276X	6b	3 (0.3)		
p.R117C	4	3 (0.3)		
CFTRdel2,3(21kb)	Deletion of exons 2 and 3	3 (0.3)		Eastern Slavic
p.C524X	10	2 (0.2)		
p.R553X	11	2 (0.2)		Swiss
c.3152delT[a]	17a	2 (0.2)	18-34-13 and 17-49-18	Greek
p.D110E[a]	4	2 (0.2)		Greek
c.4382delA	24	2 (0.2)		Mediterranean
c.1898+1G>A	intron 12	2 (0.2)		Welsh
c.2790-1G>C	intron 14b	2 (0.2)		
p.M348K	7	2 (0.2)		Mediterranean
p.E831X	13	2 (0.2)		Syrian
p.L1065F[a]	17b	2 (0.2)		Greek
p.A46D+c.3272-4A>G[a]	2 and intron 17a	1 (0.1)		Greek
c.1898+1G>T	intron 12	1 (0.1)	19-8-13	East Asian
p.D1445N[a]	24	1 (0.1)		Greek
p.D1152H	18	1 (0.1)		
c.296+1G>C[a]	intron 2	1 (0.1)		Greek
p.R75L	3	1 (0.1)		
c.3849+4A>G	intron 19	1 (0.1)		
p.R1162X	19	1 (0.1)		Multi-ethnic
p.G1069R	17b	1 (0.1)		
p.E725K[a]	13	1 (0.1)		Greek
p.F1052V	17b	1 (0.1)		
c.525delT	4	1 (0.1)		Yugoslavian
p.R347P	7	1 (0.1)		
p.R347H	7	1 (0.1)		
p.L732X	13	1 (0.1)		
p.D836Y	14a	1 (0.1)		
c.2751+2T>A[a]	intron 14a	1 (0.1)	18-32-13	Greek
p.F1074L	17b	1 (0.1)		
c.1497delGG[a]	intron 9	1 (0.1)	18-30-17	Greek
p.S813P[a]	13	1 (0.1)		Greek
c.1571delG	10	1 (0.1)		Slavic
c.2118del4	13	1 (0.1)		
c.547insTA	4	1 (0.1)		

Table 1. Continued

Mutation	Location Exon/intron	No. of chromosomes (Frequency %)	Microsatellite haplotype[b]	Ethnic prevalence[c]
p.G1244V	20	1 (0.1)		
p.Q2X; p.R3W	1	1 (0.1)		
p.R75X	3	1 (0.1)		Hispanic
c.1716+1G>A[a]	intron 10	1 (0.1)		Greek
c.441delA	4	1 (0.1)		
p.D565G[a]	12	1 (0.1)		Greek
p.R352W	7	1 (0.1)		
p.Y247X[a]	6a	1 (0.1)	17-34-13	Greek
p.S945L	15	1 (0.1)		
p.P1021A[a]	17a	1 (0.1)		Greek
p.S997F	16	1 (0.1)		
pE823X[a]	13	1 (0.1)		Greek
c.4040delA	21	1 (0.1)		
c.2752-26A>G[a]	intron 14a	1 (0.1)	17-25-13	Greek
p.L1093P	17b	1 (0.1)		
c.406-6T>C	intron 3	1 (0.1)		Mediterranean
c.457TAT>G	4	1 (0.1)		Mediterranean
c.3849+10kbC>T	intron 19	1 (0.1)		Ashkenazi Jewish
c.2418GG>T[a]	13	1 (0.1)		Greek
Unknown		79 (7.9)		
TOTAL		100		

[a]Mutations described for the first time by our group. [b]Most common haplotype in the Greek population. [c]Bobadilla et al., 2002; Estivill et al., 1997; CFGAC database. Variants are described using DNA and protein designation: intronic changes, deletions, insertions and frameshifts are reported at the cDNA level (c.) and amino acid changes at the protein level (p.)

Haplotype Analysis

The phase of the diallelic and microsatellite haplotypes was determined by family studies (typing of both parents and affected CF child). For the normal alleles a total of 27 different diallelic [most common 11117212 (16%)] and 69 microsatellite [most common 17-29/30/31-13 (19.1%) and 18-31/32-13 (15.4%)] haplotypes were found. A total of 16 diallelic and 18 microsatellite haplotypes were found for p.F508del. Haplotype 12216112 (65.2%) was the most common (in accordance with other European populations) while 25-31/32-13 (40.4%), 26-32/33-13 (15.7%) and 17/19-31-13 (8.9%) were the most common p.F508del microsatellite haplotypes. 25-31-13 may have arisen by slippage from the most common European 23-31-13 haplotype, which has a frequency of only 6.5% in the Greek population. The non-p.F508del alleles were associated with 22 diallelic (most common 12216112 – 25.7%, 11127212 – 3.7%) and 31 microsatellite haplotypes. The most common haplotype for c.621+1G>T was 23-32-13, for p.G542X 25-33/34-13, for p.N1303K 25-32-13 and for p.E822X 17-30-13 (tables 2–4).

Table 2. CFTR haplotypes for Diallelic and Microsatellite markers for 217 p.F508del Greek chromosomes

XV2C-KM19-D9-J44-GATT-M470V-T854T-TUB20	**8CA-17BTA-17BCA**
1-2-2-1-6-1-1-2 (135)	**25-31-13(26), 25-32-13(15),** 25-33-13(2), 25-27-13(1), 25-49-13(1), 25-7-13(1), 25-35-13(2), **26-32-13(10), 26-33-13(7),** 26-31-13(2), 26-34-13(2), **23-31-13(3),** 23-28-13(2), 23-32-13 (1), **24-31-13(3),** 24-33-13(2), **19-31-13(2),** 19-29-13(1), 19-32-13(2), 19-34-13(1), **17-31-13(4),** 17-30-13(2), **18-32-13(3),** 18-31-13(1), 18-29-13(1)
1-1-2-1-6-1-1-2 (20)	**25-31-13(6), 25-32-13(1),** 25-30-13(1), **26-32-13(2),** 26-31-13(1), 17-30-13(1)
1-2-1-1-6-1-1-2 (14)	**25-31-13(2), 25-32-13(1), 17-31-13(2), 19-31-13(1)**
1-2-2-1-6-1-2-2 (8)	**25-31-13(1), 25-32-13(1),** 26-34-13(1), 19-34-13(2)
2-2-2-1-6-1-1-2 (8)	**18-32-13(2), 19-31-13(2)**
1-1-1-1-6-1-1-2 (5)	**25-32-13(2), 26-32-13(1)**
1-2-2-1-6-1-1-1 (5)	**26-33-13(1),** 18-29-13(1), 23-31-13(1), 25-27-23(1)
2-1-1-1-6-1-1-2 (2)	**25-31-13(2),** 24-32-13 (1)
2-2-1-1-6-1-1-2 (2)	**25-31-13(1)**
2-1-1-2-6-1-1-2 (2)	17-24-13(1), 17-34-13(1)
2-1-2-1-6-1-1-2 (1)	**19-31-13(1)**
1-1-2-2-6-1-1-2 (1)	17-32-13 (1)
2-1-2-2-6-1-1-2 (1)	17-32-13(1)
2-2-2-1-6-1-2-2 (1)	**26-33-13(1), 19-31-13(1), 18-32-13 (2), 24-31-13(1)**
2-2-2-1-6-1-1-1 (1) 1-2-2-2-6-1-1-2 (1)	**25-31-13(1)**

Allele 1 denotes the absence of the restriction site, whereas Allele 2 denotes its presence. Numbers for microsatellites GATT, IVS8CA, IVS17BTA, and IVS17BCA are the number of repeats. Numbers in parentheses denote the number of chromosomes for the specific haplotype in p.F508del chromosomes. Most common diallelic and microsatellite haplotypes are in bold letters. Most common microsatellite haplotypes are: 25-31-13(39), 25-32-13(20), 26-32-13(15), 26-33-13 (8), 19-31-13(7), 17-31-13(6), 18-32-13(5), 24-31-13(4), 17-30-13(3), 23-31-13(3).

Male Infertility: CBAVD, Obstructive Azoospermia (OA) and Oligospermia

In the 48 males with CBAVD, a total of 22 different molecular defects were observed, with 56.2% carrying some type of molecular defect in the CFTR gene, and 27.1% being compound heterozygotes of two mutations or a mutation and the IVS8-5T allele. As expected a high frequency of p.F508del (20.8%) was found. Some of the mutations were common with CF patients (p.W496X, c.621+1G>T, c.4010delTAT, p.R1158X), while the rest except for p.L932X were missense or splicing with high residual levels of functional CFTR (p.M1I, p.R352W, c.711+3A>G, p.F1052V, p.R117C, p.D110E, p.D110H, p.D565G, c.3272-26A>G, p.I521F, p.P1021A, p.G817V, p.M952I). The presence of genomic variants with variable penetrance (p.F508C, p.R668C, p.R75Q, p.R1162L) was also noted, as well as a high frequency (11/48, 22.9%) of IVS8-5T exon 9 mis-splicing variant, compared to 3% in the controls.

Table 3. Haplotypes for diallelic and microsatellite DNA markers for CFTR mutations in Greek CF patients

Mutation	XV2C-KM19-D9-J44-GATT-M470V-T854T-TUB20	8CA-17BTA-17BCA
c.621+1G>T (26)	**1-2-2-1-6-1-1-2 (20),** 1-1-2-1-6-1-1-2 (3), 2-2-2-1-6-1-1-2(2)	**23-32-13(7),** 24-32-13(4), 23-31-13(4), 24-33-13(2), 24-8-17(1), 22-8-13(1), 18-8-17(1)
p.G542X (19)	**1-2-2-1-6-1-1-2 (10),** 2-2-2-1-6-1-1-2 (3), 1-1-1-6-1-1-2 (2), 2-2-1-1-6-1-1-2 (1), 1-2-1-1-6-1-1-2 (2), 1-1-1-2-1-6-1-1-2 (1)	**25-34-13(3), 24-33-13(3),** 25-31-13(1), 19-35-13(1), 26-32-13(1), 17-31-13(1)
p.N1303K (14)	**1-2-2-1-6-1-1-2 (5), 1-1-2-1-6-1-1-2 (5),** 2-2-2-1-6-1-1-2 (2), 1-1-1-2-7-1-1-2(1), 1-2-1-1-6-1-1-2 (1)	26-32-13(2), 25-32-13(2), 26-33-13(1), 25-7-13(1), 17-8-13(1), 25-28-13(1)
c.2183AA>G (7)	**2-2-2-1-7-1-1-2 (3),** 1-2-2-1-6-1-1-2 (2), 2-2-1-1-7-1-1-2(1)	17-29-13(2), 17-31-13(1)
p.E822X (10)	**1-1-1-2-7-2-1-2 (5),** 1-1-1-1-7-2-1-2 (2), 2-1-1-2-7-2-2-2 (1)	**17-30-13(7),** 17-32-13(2), 17-7-18(1)
c.2789+5G>A (3)	1-1-1-1-8-2-1-2 (1), 2-1-2-1-8-1-2-1 (1), 1-2-2-1-7-1-1-2(1)	18-8-18 (1), 18-28-18(1)
p.R1158X (6)	**2-2-2-1-6-1-1-1 (3),** 2-1-1-2-6-1-1-2 (1), 1-2-1-1-6-1-1-2(1), 2-1-1-1-7-2-1-2 (1)	**25-34-13(3),** 25-33-13(1), 17-28-13(1)
p.R1070Q (3)	1-1-2-1-6-1-1-2 (2), 1-1-1-2-7-1-1-2(1)	18-35-13(1), 19-32-13(1), 17-31-13(1)
p.G85E (5)	2-2-2-1-8-2-2-1 (2), 1-2-2-1-8-2-2-1(1), 1-1-1-2-8-2-1-2(1)	**18-8-18(3),** 18-24-13(1)
c.3272-26A>G (4)	2-2-2-1-?-1-2-1(2), 1-1-1-?-2-2-1 (1)	**18-8-18(3)**
p.I507del (4)	1-2-2-1-6-1-1-2 (1), 2-2-2-1-6-1-1-2 (2), 1-2-1-1-6-1-1-2(1)	26-32-13(1), 20-32-13(1), 24-31-13(1), 18-31-13(1)
p.W1282X (2)	1-2-2-1-6-1-2-1(2)	18-8-17(1)
p.C276X (1)	1-2-1-1-?-?-1-2	18-47-13
c.1898+1G>T (1)	1-1-1-2-?-?-1-2	19-8-13
c.3849+10kbC>T (1)	1-2-2-1-7-1-1-2	17-31-13
p.W496X (1)	2-2-2-2-?-1-1-2	18-32-13
p.A46D (1)	1-2-2-1-7-1-1-2	25-33-13
c.3152delT (2)	2-1-1-2-7-1-1-2 (1), 2-2-2-1-7-1-2-2 (1)	18-34-13, 17-49-18
c.2752-26A>G (1)		17-25-13
p.Y247X (1)	2-1-1-2-6-1-1-2	17-34-13
c.1497delGG (1)	1-2-2-1-7-1-2-1	18-30-17
p.L1065F (1)	1-2-2-2-6-2-1-2	
p.K162E (1)		18-30-13
c.2751+2T>A (1)	1-1-1-2-7-2-1-2	18-32-13

Allele 1 denotes absence of restriction site, whereas allele 2 denotes its presence. Numbers for microsatellites GATT, IVS8CA, IVS17BTA and IVS17BTA are the number of repeats. Numbers in parentheses denote the number of chromosomes with the specific haplotype. Haplotypes in bold denote the most common haplotype for the mutation. DNA and protein designation: intronic changes, deletions, insertions and frameshifts are reported at the cDNA level (c.) and amino acid changes at the protein level (p.)

Table 4. Allele Frequencies for the intron 8 CA repeat in normal and CF chromosomes

FREQUENCY OF No of repeats		**Normal (n=136)**		**p.F508del (n=146)**		**c.621+1G>T (n=20)**		**p.G542X (n=10)**		**p.N1303K (n=8)**		**p.E822X (n=10)**		**p.R1158X (n=5)**	
8CA	17BTA	8CA	17BTA	8CA	17BTA	8CA	17BTA	8CA	17BTA	8CA	17BTA	8CA	17BTA	8CA	17BTA
13	7	0.007	0.015		0.007						0.125		0.1		
16	***8***	0.022	***0.11***				0.2				0.125				
17	9	**0.294**	0.074	0.089				0.1		0.125		**1**		0.2	
18	18	**0.397**	0.007	0.068		0.05									
19	20	0.132	0.007	0.089				0.1							
20	22	0.015	0.007												
22	23	0.015	0.007			0.05									
23	24	0.007		0.048	0.007	***0.55***									
24	25	0.044	0.007	0.048		***0.35***		***0.3***							
25	26	0.029	0.007	**0.466**				**0.4**		**0.5**				0.8	
26	27	0.007		**0.192**	0.014			0.1		**0.375**					
	28		0.015		0.014						0.125				0.2
	29		0.081		0.02										
	30		***0.103***		0.027								***0.7***		
	31		**0.22**		**0.438**		**0.2**		0.2						
	32		**0.147**		**0.315**		**0.55**		0.1		**0.5**		0.2		
	33		0.066		0.089		0.2		***0.3***		0.125				0.2
	34		0.029		0.048				***0.3***						***0.6***
	35		0.015		0.014				0.1						
	36		0.015												
	38		0.015												
	45		0.029												
	47		0.007												
	48		0.007												
	49				0.007										
	55		0.007												

Note- numbers of chromosomes studied appear in parentheses; numbers in bold: most common haplotypes. DNA and protein designation: intronic changes, deletions, insertions and frameshifts are reported at the cDNA level (c.) and amino acid changes at the protein level (p.)

Of the 40 OA males 27.5% had at least 1 CF allele, and 3 carried a severe mutation (p.F508del, p.W496X, and p.W1282X) and the IVS8-5T allele in trans or genomic variants with variable penetrance. Of the 46 males with severe oligospermia, 13% were found to be carriers of a CFTR mutation/variant and 2.2% carriers of the IVS8-5T allele (table 5).

Table 5. Summary of CFTR mutation screening on male infertility patients

CBAVD Patients (n=48)		**Obstructive Azoospermia(n=40)**	**Oligospermia (n=46)**
2 mutations	**1 mutation AND/OR IVS8-5T**		
p.F508del/p.M1I and IVS8-5T	p.F508del(4), and IVS8-5T(3)	p.W496X/p.I506V(1)	p.F508del(2)
p.F508del/c.711+3A>G	p.D565G(3)	p.F508del and IVS8-5T(1)	p.G542X(1)
p.F508del/p.F508C	c.3272-26A>G(1)	c.2790-8C>G(1)	p.C491S(1)
p.F508del/p.D110E	p.L932X and IVS8-5T(1)	p.S549N(1)	p.F1052V(1)
p.F508del/p.R352W	c.711+3A>G and IVS8-5T(1)	p.W1282X and IVS8-5T(1)	c.2752-6T>C(1)
p.F508del/p.M952I	c.4010delTAT(1)	p.F1052V(1)	IVS8-5T(1)
c.621+1G>T/p.R117C	p.I521F(1)	p.D110H(1)	
p.F1052V/p.F1052V	p.R1158X(1)	c.406-6T>C(1)	
	p.P1021A(1)	p.R75Q(2)	
	p.D110H(1)	IVS8-5T(1)	
	p.G817V(1)		
	c.3659delC(1)		
	IVS8-5T(3)		
Total: 8	**Total: 20**	**Total:11**	**Total: 7**

Numbers in parentheses denote number of chromosomes. DNA and protein designation: intronic changes, deletions, insertions and frameshifts are reported at the cDNA level (c.) and amino acid changes at the protein level (p.)

Heterozygote Screening

Mutation analysis on 900 individuals identified a total of 188 heterozygotes. More specifically 58/91 heterozygotes (63.7%) were identified amongst the siblings of CF index cases, 115/285 (40.4%) amongst the CF family relatives (R) and 15/174 (8.6%) amongst the general population individuals (GP). Amongst the 175 couples screened, 14 had a CF partner (GP/CF), 106 had a partner with family history of CF (GP/R), 23 had a male partner with infertility (GP/CBAVD) and in 32 couples both were general population subjects (GP/GP). Of the 175 couples, 23 were already pregnant upon request for CFTR screening, and six were found requiring prenatal diagnosis [1 GP/R (0.9%), 2 GP/GP (6.25%), 2 GP/CBAVD (8.7%) and 1 GP/CF (7.1%)].

Prenatal Diagnosis

For the 167 prenatal diagnoses performed a total of 57 different genotypes (of which 13 from the IVF couples), were assessed. Direct analysis was possible in all but four cases for which only one parental mutation was known. In these cases microsatellite analysis in addition to testing for the one parental mutation was conclusive for three and not possible for one. This fetus was diagnosed as having the one parental allele through mutation analysis, and the microsatellite sites were not informative. In 46 prenatal diagnoses the fetuses were diagnosed as normal, in 80 as heterozygotes and 41 as affected. The results are in agreement with those expected for an autosomal recessive disease. The outcome of all cases that were diagnosed as being unaffected by CF was a normal child. In two cases where one of the parents had two mutations in cis, p.R1070Q+p.S466X and p.Q2X+p.R3W, respectively, fetal DNA was tested for one of the mutations of the complex allele (p.R1070Q and p.Q2X, respectively). Similarly in four couples with one parent compound heterozygote but asymptomatic (Genotypes: p.F508del/p.K162E and p.R1158X/p.R1070Q) and two mono-symptomatic CBAVD parents (Genotypes: p.F1052V/p.F1052V and p.F508del/p.M952I), fetal DNA was tested for 3 parental mutations. If the embryo combined one severe and one mild parental mutation, genetic counselling on the clinical phenotype is difficult. Such a case arose in two prenatal diagnoses involving IVF couples. The fetuses were diagnosed as homozygous with genotypes p.F1052V/p.M348K and p.G565G/p.406-6T>C respectively. Both couples were counselled on the expected mildness of the phenotype, and both decided to carry the pregnancies to term.

Table 6. Summary of echogenic bowel families tested and CF genotype results

Both parents and/or fetus negative	24	
One parent carrier- fetus negative	2	p.I148T, p.F508del
One parent carrier- fetus not tested	5	c.621+3A>G, p.R75Q and 3 with p.F508del
One parent carrier- fetus carrier	7	c.621+1G>T, p.I148T and 5 with p.F508del
Both parents carriers- fetus affected	4	p.F508del/p.L732X, p.F508del/c.574delA, p.F508del/p.F508del, p.F508del/p.G542X
Fetus only tested- negative	29	
Fetus only tested- carrier	8	p.E822X; c.2789+5G>A; p.R297Q; and 5 p.F508del
Couple only tested–both negative	31	
TOTAL	110	4 homozygotes AND 15 with 1 detectable mutation

Prenatal Diagnosis due to Fetal Bowel Echogenicity

Seventy-four AF samples were tested for CF mutations due to bowel hyperechogenicity in the fetus, detected during a routine second trimester ultrasound examination. Homozygosity was detected in 4 cases (3.6%), with the following severe disease indicating genotypes: p.F508del/p.L732X, p.F508del/c.574delA, p.F508del/p.F508del, and p.F508del/p.G542X. Fifteen of the (13.6%) had one detectable mutation, 10 with p.F508del, and 1 each for c.621+1G>T, p.I148T, c.2789+5G>A, p.R297Q, p.E822X. An additional finding was the presence of particular polymorphisms: 5/110 instances of 1716G/A (p.E528E), 2 of c.2752-15G/C, 2 of p.R1162L and 1 of c.4029A/G. The significance of these findings can not be readily explained (table 6).

Conclusion

Molecular Heterogeneity of CF in Greece

The most common CFTR mutations worldwide, showing frequencies >1%, are the following: p.F508del, p.G542X, p.N1303K, p.G551D and p.W1282X. Another 19 mutations are found in 0.1–1% of cases showing considerable variation for different populations, with the majority being restricted to particular geographic areas. Mutation p.F508del shows a North-to-South Europe decreasing frequency gradient (30–90%) while the percentage of non-p.F508del chromosomes associated with CF increases [CFGAC: http://www.genet.sickkids.on.ca/cftr; Estivill et al., 1997; Bobadilla et al., 2002]. The Mediterranean area has the highest CFTR heterogeneity, making genetic diagnosis by mutation analysis especially difficult [Claustres et al., 1996; Chillon et al., 1994; Casals et al., 1996; Kanavakis et al., 2003; Mendes et al., 2003; Castaldo et al., 2004; des Georges et al., 2004; Radivojevic et al., 2004]. For the Spanish CF patients, 73 different mutations have been identified accounting for 87% of CF alleles with only 10 of them having a frequency higher than 1% and for the French population 105 mutations account for 86% of CF chromosomes [Casals et al., 1996; Gasparini et al., 1993; des Georges et al., 2004]. Studies of North-eastern Italian CF patients identified 62 mutations accounting for 73.8% of detectable CF alleles while for the Celtic population from Brittany, France, 19 mutations account for 98% of CF chromosomes [Gasparini et al., 1993; Ferec et al., 1992; Des Georges et al., 2004].

DGGE analysis of the 27 exons and neighboring intronic regions of the CFTR gene revealed the presence of 83 different mutations covering 92% of CF alleles, proving that the Greek population has one of the highest rates of CFTR mutation heterogeneity. Twenty of the characterized mutations are apparently specific to the Greek population and the majority are found in one or two alleles with the exception of p.E822X which has achieved a high frequency in Central Greece. The list of mutations found in our population is shown in table 1. Three of these (p.F508del, p.G542X and p.N1303K) are common worldwide, the rest with the exception of p.E822X, are common in the Mediterranean area, some achieving relatively high frequencies in wide geographic areas and others being common in small regions [CFGAC: http://www.genet.sickkids.on.ca/cftr; Estivill et al., 1997; Bobadilla et al., 2002]. The wide distribution of p.F508del, p.G542X and p.N1303K suggests an ancient origin for these mutations [Morral et al., 1994]. Mutation p.G542X is common in the Mediterranean, with a mean frequency of 4.3%, while p.N1303K is found in most of the western Mediterranean countries with highest frequency in Tunisia (17.2%). The other common mutations such as c.621+1G>T, considered a Celtic mutation, c.2789+5G>A and c.2183AA>G, considered Southern European, have achieved relatively high frequencies in the south-eastern Mediterranean region. Mutation p.R1158X is very rare and is found in southern Italy with comparable frequencies to those found in Greece with only two alleles described in France and one in Spain; therefore, it must have an origin and diffusion in the specific European region [Claustres et al., 1996; Tzetis et al., 1997; Castaldo et al., 1999; Castaldo et al., 2004; Chillon et al., 1994]. Mutation p.E822X has only been found in the Greek population and is confined to a specific region of Central Greece. Sixteen mutations show frequencies between 0.4 and 1.1% (table 1). Of these c.1677delTA and p.R1070Q have reached high frequencies in Bulgaria and Turkey, c.3272-26A>G in Portugal, France and Belgium (with frequencies ranging from 1% to 2.8%) and c.3120+1G>A is considered an African mutation, reaching high frequencies amongst the African–Americans (13.9%) [Dork et al., 1998; Bobadilla et al., 2002]. Mutation p.R334W is common in the Eastern Mediterranean, reaching a very high frequency in Crete. p.W1282X has the highest frequency in Israel, being the most common mutation and c.621+3A>G also found in the French population, initially reported as a polymorphism was later proven to be a splicing mutation [CFGAC database http://www.genet.sickkids.on.ca/cftr; Estivill et al., 1997; Amaral et al., 2001; Dork et al., 1998; Tzetis et al., 2001b].

Haplotype Analysis

The normal chromosomes were found associated with a total of 11 alleles for the IVS8CA site (17 and 18 being the commonest) and 23 for the IVS17BTA site (30, 31 and 32 being the commonest). The IVS17BCA site showed the lowest variability with only six alleles of which 13 and 17 had the highest frequency. Mutation p.F508del associated with seven alleles for the IVS8CA site (25 and 26 most common), 12 IVS17BTA alleles (31 and 32 most common) and only allele 13 for the IVS17BCA site. The other common mutations showed lower allele variability for both the IVS8CA and IVS17BTA sites (tables 2–4). Strong linkage disequilibrium was observed for p.F508del and one diallelic haplotype (12216112 –65.2%), which accounts for only 0.02% of normal Greek chromosomes. The other two most frequent disease causing mutations such as c.621+1G>T, p.G542X and p.N1303K also reside on the typical p.F508del haplotype. This specific diallelic haplotype is commonly associated with p.F508del and common mutations in other populations supporting the theory of an ancient origin of these mutations [Dork et al., 1992; Morral et al., 1993; 1994; 1996]. Microsatellite haplotype 25-31-13, which may have arisen by slippage from the most common European haplotype 23-31-13, was found most commonly associated with p.F508del, while p.G542X (25-33/34-13) and p.N1303K (25-32-13) have the 8CA-25 repeat allele in common with p.F508del. The most common haplotype for mutation c.621+1G>T was 23-32-13 (tables 2 and 3). The observed higher heterogeneity for the p.F508del diallelic haplotype background and the difference in the major microsatellite haplotype from Northern European populations supports the theory of a longer history of p.F508del in Southern Europe which would have allowed time for recombination with other haplotypes [Claustres et al., 1996; Morral et al., 1994]. Mutation p.R1158X, which is specific to the Southern Mediterranean region, is on a different diallelic and microsatellite framework than that observed for Italy and France, suggesting a recurrent origin or several recombination events [Castaldo et al., 1999]. The microsatellite haplotype was 16-7-17 for Italian CF chromosomes and 16-7/45-17/13 for the two French alleles being different from the Greek 25/17-28/33/34-13 (25-34-13; 3 alleles). The mutation could have been introduced to Southern Italy from ancient Greeks. The rare and specific Greek mutations are each associated with specific diallelic and microsatellite markers, which agrees with the theory that these mutations have a fairly recent origin. This especially applies to mutation p.E822X which is exclusively associated with a microsatellite framework (17-30-13) which is common in the normal chromosomes and with the commonest diallelic haplotype (11117212) for the normal alleles. The analysis of haplotypes and mutations in the Greek population adds to the information on CFTR gene mutations in Europe. This study also confirms that polymorphism analysis is a valid procedure to asses the genetic heterogeneity of a population and to evaluate the origin of a mutation. The high mutational heterogeneity and haplotypic variability shown in the Greek population is consistent with an earlier appearance of CFTR mutations in southern Europe, which would have allowed time for haplotypic recombination and slippage at microsatellites to occur [Claustres et al., 1996; Morral et al., 1996; Bertrapetit et al., 1996; Angelicheva et al., 1997].

Monosymtomatic CFTR

Male Infertility: CBAVD, Obstructive Azoospermia (OA) and Oligospermia

CF-related infertility in men is primarily due to obstruction or absence of the vas deferens [Welsh et al., 2001] though a defect in spermatogenesis may be a contributing factor, as decreased numbers of mature spermatids and the presence of many malformed late spermatids are also found in CF [Trezise et al., 1993].

The relationship between CBAVD and CF gene mutations has previously been demonstrated [Chillon et al., 1995; Costes et al., 1995; Kanavakis et al., 1998; Claustres et al., 2000; Claustres, 2004]. In the majority of cases two CFTR mutations (including the IVS8-5T allele) were present in approximately 47% and one mutation in 24% of CBAVD patients. The IVS8-5T allele is the most commonly found CFTR defect in CBAVD males [Chillon et al., 1995; Estivill et al., 1996; Kanavakis et al., 1998; Claustres et al., 2000; Groman et al., 2004]. Generally one “severe” allele is combined with one “mild” or variant allele, such that the mild allele appears to dominate and cause a milder phenotype. These milder alleles may produce CFTR protein with low levels of expression (8-10% of normal) and thus cause disease in the vas deferens tissue and not in other organs affected by cystic fibrosis. Increased frequency of CF mutations in healthy men with reduced sperm quality and also in men with obstructive azoospermia with the presence of the vas deferens has also been reported [Van der Ven et al., 1996; Meschede et al., 1997; Kanavakis et al., 1998].

Our results show that of the 48 CBAVD males, 27.1% were compound heterozygotes, the majority with p.F508del (20.8% vs 54.5% for classic CF patients) and a mild mutation/variant in trans including the IVS8-5T allele, while 25% had only one mutation. The frequency of the IVS8-5T allele among CBAVD males was 22.9% compared to 3% in the general population. In this group of patients a total of 22 different defects were observed with the majority of mutations except for p.R352W, p.M952I, p.L932X, c.4010delTAT, p.G817V, c.3659delC, c.2790-8C>G, p.S549N, p.C491S, c.2752-6T>C and p.M1I also found in our cohort of CF patients (tables 1 and 5). In only four of the eight compound heterozygote CBAVD patients the sweat test was >60mEq/l. Furthermore three of the 48 CBAVD men also presented with recurrent bronchitis or chronic cough.

Of the 40 men with OA, 27.5% had at least one CF allele and three carried a severe mutation also found in our cohort of CF patients (p.F508del, p.W496X, p.W1282X) and the IVS8-5T allele in trans or genomic variants with variable penetrance. The frequency of the IVS8-5T allele was 7.5%. The sweat test was borderline positive (40-60mEq/l) in three patients. The high frequency of CFTR mutations in males with OA (27.5%) suggests that the CFTR protein may play a critical role in spermatogenesis or sperm maturation that is independent of its function in the development of epithelial ducts such as the vas deferens [Trezise et al., 1993; Van der ven et al., 1996; Meschende et al., 1997].

Of the 46 males with severe oligospermia, 13% were found to be carriers of a CFTR mutation/variant and 2.2% carriers of the IVS8-5T allele (table 5).

In conclusion CFTR mutations may represent one of the most common abnormalities associated with male infertility, especially with CBAVD (CFTR mutation frequency 56.2%) but also in OA (27.5%). These results are relevant for couples treated by in-vitro fertilization (IVF) and intracytoplasmic sperm injection (ICSI). Patients with male infertility and their partners should be extensively tested by scanning techniques such as DGGE covering exons that harbour at least 85% of CFTR mutations of the specific ethnic group. Because of the incomplete penetrance of mild mutations for some combinations of CFTR genes in these couples (rare missense and severe mutation), counselling on the phenotypic consequences for their children is very difficult. We have encountered such combinations in two couples (with the following genotypes: p.F1052V/p.M384K and p.G565G/c.406T>C), which are discussed further below in this chapter. Finally, patients with CBAVD who have a positive sweat test and are compound heterozygotes for two CFTR mutations or one mutation/and the IVS8-5T allele probably warrant further clinical follow-up for potential late complications of CF, especially advice on persistent pulmonary symptoms.

Heterozygote Screening

For recessive conditions such as CF the goal of genetic screening is to recognize asymptomatic heterozygous carriers and to aid the process of reproductive decision making. Two distinct types of genetic screening strategy have been proposed. One offers testing to the population at large; the other targets high-risk groups identified through a family history or a specific clinical phenotype [Super et al., 1994; Brock 1995; Mennuti 2001; Palomaki et al., 2004]. Carrier screening for the population at large is not currently possible, we therefore concentrated on testing individuals belonging to a high-risk group including siblings of index cases, CF family relatives (R) and their partners (when available) and both members of couples seeking IVF due to male infertility. A common reason for performing IVF and intracytoplasmic sperm injection (ICSI) is congenital bilateral absence of vas deferens (CBAVD). Because of the high frequency of CFTR mutations found in infertile males with CBAVD [Kanavakis et al., 1998; and results in this chapter], the male partner needs to be screened and couple screening is recommend in such cases. The remaining carrier risk, after negative screening for 85% of detectable CF alleles, is 1/161. Couples in which one partner is a definite carrier and the other has a negative mutation analysis have their risk of having an affected child reduced to 1/644 from 1/100, and in cases when neither partner tests positive the risk is 1/103324. For most couples this could be reassuring information. Through such a screening program on 900 individuals a total of 188 heterozygotes were identified. As expected a high number of heterozygotes were found amongst the 91 siblings of CF index cases (63.7%) and amongst the CF family relatives (R) (40.4%). The carrier frequency amongst the general population individuals was 8.6% (15/174). Six couples where both members were carriers were identified amongst the 175 couples screened [1GP/R (0.9%), 2GP/GP (6.25%), 2GP/CBAVD (8.7%) and 1GP/CF (7.1%)]. Prenatal diagnosis was performed in these cases including two twin pregnancies conceived through ICSI, because the male partner had CBAVD. The results of the pregnancies were one homozygous, four heterozygous and three normal infants. While diagnostic methods are not automatic and robust enough for heterogeneous CF populations, the diagnostic strategy of DGGE screening of 15 CFTR exons (accounting for 86% of CF alleles) in combination with haplotype association of certain mutations is presently an appropriate tool for screening the Greek population. This strategy also provides rapid, accurate and reliable prenatal diagnosis for the majority of couples.

Prenatal Diagnosis

The characterization of mutations and their frequencies in a specific population is a prerequisite for setting up screening and prenatal diagnosis programs. For the highly heterogeneous Greek population a strategy of DGGE and sequencing resulted in the characterization of 92% of CF alleles. In the case of families with a one in four risk, if the causative mutation(s) have been identified in both parents before pregnancy, DGGE analysis permits the determination of presence or absence of the mutation(s) in the chorionic villus or amniotic fluid sample within 2 days. The additional use of the polymorphic microsatellite markers permits prenatal diagnosis in cases where only one parental mutation is known and also checks for maternal contamination and confirms paternity of the fetal DNA sample. For the 176 prenatal diagnoses performed a combination of direct mutation and microsatellite analysis was conclusive in all but one case. Microsatellite analysis is performed even in those cases where the causative mutation(s) have been identified in a given family. The extremely sensitive technique of fluorescent PCR will reveal the presence of even minute quantities of maternal contamination in the fetal sample. We have not observed any cases of maternal contamination in any of the fetal samples we have so far tested. A factor that complicates CFTR mutation screening and prenatal diagnosis is the presence of complex alleles. We observed two such cases in which parental alleles had two mutations in cis (p.R1070Q+p.S446X and p.Q2X+p.R3W, respectively). In such cases careful analysis must precede the prenatal diagnosis for the confirmation of the cis positioning of the two alleles. In the course of genotyping our CF patient population we observed two cases where one parent, without clinical symptoms of CF, is a compound heterozygotes for two CF mutations (p.F508del/p.K162E and p.R1158X/p.R1070Q). In such cases the fetal DNA must be tested for three parental mutations. In the course of screening for male infertility we identified two couples requiring prenatal diagnosis where the CBAVD males were compound heterozygotes (p.F1052V/p.F1052V and p.F508del/p.M952I). In these cases fetal DNA was also tested for three parental mutations. When the genotype of the embryo combines one parental severe mutation and one mild mutation, genetic counselling on the clinical phenotype of the embryo is difficult. In two IVF couples the fetuses were diagnosed with the following genotypes: p.F1052V/p.M384K and p.G565G/c.406T>C. The couples after being counselled on the expected mildness of the phenotype, decided to carry the pregnancies to term. These cases should be elucidated when modifying factors (possibly other genes) affecting the clinical phenotype are identified.

Prenatal Diagnosis due to Echogenic Bowel

For pregnancies presented with echogenic bowel (EB) we have tested both parental and fetal DNA in the majority of cases. A negative screening of about 62 mutations (86% of detectable CF alleles) reduces the probability of CF diagnosis when both parents have been excluded as carriers. According to the currently available literature when no CF mutations are identified the risk that the fetus is affected with CF is between 1/453 and 1/1516, while a fetus with one CF mutation has a 21–47% probability of being affected with CF. Direct fetal studies should be performed to additionally avoid cases of non-paternity which could lead to false–negative diagnosis and testing for chromosomal abnormalities [Bosco et al., 1999; Monaghan et al., 1999; Strocker et al., 2000; Sepuvelda et al., 2000]. We identified four homozygous fetuses (3.6%) and fifteen (13.6%) with a single CF mutation (table 6). Our data indicated that the prior risk of CF in a fetus with echogenic bowel for our population is 3.6% (1/28 births, ~90 times higher than prevalence in the general population) and the remaining risk of a heterozygous fetus having CF is 17.23% [Bosco et al., 1999]. The 4 couples with affected fetuses opted for pregnancy termination. Parents were given genetic consultation of ~17% remaining risk of CF for the 15/110 fetuses with one detected mutation. Thirteen continued to term and resulted in apparently healthy infants after clinical assessment. One family opted for pregnancy termination and one pregnancy is still ongoing.

Fetal echogenic bowel is an ultrasound marker associated with various fetal complications one of which is Cystic Fibrosis. A difficult situation arises when the fetus carries one identifiable CF mutation. Assuming a prior risk of CF in EB of 3.6% as calculated from our cohort of cases, the residual risk for the fetus being affected can be calculated to be ~17%. This is an important finding that will help in the correct counseling of these couples.

The expanded research into the CFTR gene and classic and monosymptomatic forms of CF adds to the challenges of appropriate testing. Knowledge of certain CFTR mutations may indicate a severe or mild phenotype, however predicting the severity of the disease from the genotype will not be possible until additional factors affecting disease severity (environmental and genetic) are elucidated. Information that will emerge during the next years will most probably open new ways for the development of mutation specific therapies. Finally there is now great hope for an ultimate cure through gene replacement therapy although important technical hurdles need to be overcome before its application on a routine basis.

Acknowledgements

The authors wish to thank Konstantina Pouliou for technical assistance. This work was partially funded by the Research Institute for the Study of Genetic and Malignant Disorders in Childhood.

References

Ahmed N, Corey M, Forstner G et al. molecular consequences of cystic fibrosis tramsmembrane regulator (CFTR) gene mutations in the exocrine pancreas. *Gut* 2003: 52: 1159-1164.

Amaral MD, Pancheco P, Beck S et al. Cystic fibrosis patients with the 3272–26A>G splicing mutation have milder disease than F508del homozygotes: a large European study. *J Med Genet* 2001: 38: 777–782.

Angelicheva D, Calafell F, Savov A et al. Cystic fibrosis mutations and associated haplotypes in Bulgaria – a comparative population genetic study. *Hum Genet* 1997: 99: 513–520.

Bobadilla JL, Macek MJr, Fine JP, Farrell PM. Cystic Fibrosis: A Worldwide analysis of CFTR mutations- Correlation with incidence data and application to screening. *Hum Mutat* 2002: 19: 575-606.

Bonizzato A, Bisceglia L, Marigo C et al. Analysis of the complete coding region of the CFTR gene in a cohort of CF patients from north-eastern Italy: Identification of 90% of the mutations. *Hum Genet* 1995: 95: 397–402.

Bertrapetit J, Calafell F. Genetic and geographical variabilityin cystic fibrosis: evolutionary considerations. *Ciba Found Symp* 1996: 197: 97–118.

Bosco AF, Norton ME, Lieberman E. Predicting the risk of cystic fibrosis with echogenic fetal bowel and one cystic fibrosis mutation. *Obstet Gynecol* 1999: 94: 1020–1023.

Brock DJH. Heterozygote screening for cystic fibrosis. *Eur J Hum Genet* 1995: 3: 2–13.

Castaldo G, Fuccio A, Cazeneuve C et al. Detection of five rare cystic fibrosis mutations peculiar to southern Italy. Implications in screening for the disease and phenotype characterization for patients with homozygote mutations. *Clin Chem* 1999: 47: 929–931.

Castaldo G, Polizzi A, Tomaiuolo R et al. Comprehensive Cystic Fibrosis mutation epidemiology and haplotype characterization in a Southern Italian Population. *Ann Hum Genet* 2004: 69: 15-24.

Casals T, Gimenez J, Ramos M, Nunes V, Estivill X. Prenatal diagnosis of cystic fibrosis in a highly heterogeneous population. *Prenat Diagn* 1996: 16: 215–222.

Casals T, De-Gracia J, Gallego M et al. Bronchiectais in adult patients : an expression of heterozygosity for CFTR gene mutations ? *Clin Genet* 2004: 65: 490-495.

Chevalier-Porst F, Bonardot AM, Gilly R, Chazalette JP, Mathieu M, Bozon D. Mutation analysis in 600 French cystic fibrosis patients. *J Med Genet* 1994: 31: 541–544.

Chillon M, Casals T, Gimenez J, Nunes V, Estivill X. Analysis of the CFTR gene confirms the high genetic heterogeneity of the Spanish population: 43 mutations account for only 78% of CF chromosomes. *Hum Genet* 1994: 93: 447–451.

Chillon M, Casals T, Gimenez J, Nunes V, Estivill X. Analysis of the CFTR gene in the Spanish population: SSCP screening for 60 known mutations and identification of four new mutations (Q30X, A120T, 1812–1G>A, and 3667del4). *Hum Mutat* 1994: 3: 223–230.

Chillon M, Casals T, Mercier B et al. Mutations in the cystic fibrosis gene in patients with congenital absence of the vas deferens. *N Engl J Med* 1995 : 332 :1475-1480.

Claustres M, Desgeorge M, Moine P, Morral N, Estivill X. CFTR haplotypic variability for normal and mutant genes in cystic fibrosis families from southern France. *Hum Genet* 1996: 98: 336–344.

Claustres M, Guittard C, Bozon D et al. Spectrum of CFTR mutations in cystic fibrosis and in congenital absence of the vas deferens in France. *Hum Mut* 2000: 16: 143-156.

Claustres M. Symposium: Genetic aspects of male (in)fertility. Molecular pathology of the CFTR locus in male infertility. *Reprod BioMedicine Online* 2005: 10: 14-41.

Cohn JA. Reduced CFTR function and the pathobiology of Idiopathic Pancreatitis. *J Clin Gastroenterol* 2005: S70-S77.

Costes B, Fanen P, Goossens M, Ghanem N. A rapid, efficient, and sensitive assay for simultaneous detection of multiple cystic fibrosis mutations. *Hum Mutat* 1993: 2: 185–191.

Costes B, Girodon E, Ghanem N et al. Frequent occurrence of the CFTR intron 8 (TG)n5T allele in men with congenital bilateral absence of the vas deferens. *Eur J Hum Genet* 1995: 3: 285-293.

Cystic Fibrosis Genetic Analysis Consortium- CFGAC. The CFTR mutation database http://www.genet.sickkids.on.ca/cftr.

Des Georges M, Guittard C, Altieri JP, Templin C, Sarles J, Sarda P, Claustres M. High heterogeneity of CFTR mutations and unexpected low incidence of cystic fibrosis in the Mediterranean France. *J Cyst Fibros* 2004: 3: 265-272.

Dork T, Neumann T, Wulbrand U et al. Intra- and extragenic marker haplotypes of CFTR mutation in cystic fibrosis families. *Hum Genet* 1992: 88: 417–425.

Dork T, El-Harith EH, Struhrmann M et al. Evidence for a common ethnic origin of cystic fibrosis mutation 3120+1G>A in diverse populations. *Am J Hum Genet* 1998: 63: 652–662.

Estivill X. Complexity in monogenic disease. *Nat Genet* 1996:12: 348-350.

Estivill X, Bancells C, Ramos C. Geographic distribution and regional origin of 272 cystic fibrosis mutations in European populations. The Biomed CF Mutation Analysis Consortium. *Hum Mutat* 1997: 10: 135–154.

Fanen P, Ghanem N, Vidaud M et al. Molecular characterization of cystic fibrosis: 16 novel mutations identified by analysis of the whole cystic fibrosis transmembrane conductance regulator (CFTR) coding regions and splice site junctions. *Genomics* 1992: 13: 770–776.

Feldman GL, Monaghan KG. Prenatal diagnosis of cystic fibrosis. In: Milunski A, ed. Genetic Disorders and the Fetus. Diagnosis, Prevention and Treatment, 4th edn. Baltimore: The John Hopkins University Press, 1998: 506–528.

Ferec C, Audrezet MP, Mercier B et al. Detection of over 98% cystic fibrosis mutations in a Celtic population. *Nat Genet* 1992: 3: 188–191.

Friedman KJ, Highsmith WEJr, Silverman LM. Detecting multiple cystic fibrosis mutations by polymerase chain reaction-mediated site-directed mutagenesis. *Clin Chem* 1991:37:753-5.

Friedman KJ, Heim R, Knowles MR, Silverman LM. Rapid characterization of the variable length polythymidine tract in the cystic fibrosis (CFTR) gene: Association of the 5Tallele with selected CFTR mutations and its incidence in atypical sinopulmonary disease. *Hum Mutat* 1997: 10: 108–115.

Gasparini P, Marigo C, Bisceglia G et al. Screening of 62 mutations in a cohort of cystic fibrosis patients from north eastern Italy: their incidence and clinical features of defined genotypes. *Hum Mutat* 1993: 2: 389–394.

Groman JD, hefferon TW, Casals T et al. Variation in a repeat sequence determines whether a common variant of the Cystic Fibrosis Transmembrane Conductance Regulator gene is pathogenic or benign. *Am J Hum Genet* 2004: 74: 176-179.

Kanavakis E, Tzetis M, Antoniadi Th, Pistofidis G, Milligos S, Kattamis C. Cystic fibrosis mutation screening in CBAVD patients and men with obstructive azoospermia or severe oligospermia. *Mol Hum Reprod* 1998: 4: 333–337.

Kanavakis E, Efthymiadou A, Strofalis S, Doudounakis S, Traeger-Synodinos J, Tzetis M. Cystic Fibrosis in Greece: molecular diagnosis, haplotypes, prenatal diagnosis and carrier identification amongst high-risk individuals. *Clin Genet* 2003: 63: 400-409.

Kerem B, Rommens JM, Buchanan JA et al. Identification of the cystic fibrosis gene: genetic analysis. *Science* 1989: 245: 1073–1080.

Mendes F, Rosa M' ,Dragomir A, et al. Unusually common cystic fibrosis mutation in Portugal encodes a misprocessed protein. *Biochem Biophys Res Commun* 2003: 311:665-71.

Mennuti MT. Lights! Camera! Action! *Obstet Gynecol* 2001: 98: 539–541.

Meschende D, Dworniczak B, Behre HM et al. CFTR gene mutations in men with bilateral ejaculatory-duct obstruction and anomalies of the seminal vesicles. *Am J Hum Genet* 1997: 61: 1200-1202.

Monaghan KG, Feldman GL. The risk of prenatal diagnosis with prenatally detected echogenic bowel in an ethnically and racially diverse North American Population. *Prenat Diagn* 1999: 19: 604–609.

Morral N, Estivill X. Multiplex PCR amplification of three microsatellites within the CFTR gene. *Genomics* 1992: 13: 1362–1364.

Morral N, Nunes V, Casals Tet al. Microsatellite haplotypes for cystic fibrosis: mutation frameworks and evolutionary tracers. *Hum Mol Genet* 1993: 2: 1015–1022.

Morral N, Bertranpetit J, Estivill X et al. The origin of the major cystic fibrosis mutation (DF508) in European populations. *Nat Genet* 1994: 7: 169–175.

Morral N, Dork T, Llevadot R et al. Haplotype analysis of 94 cystic fibrosis mutations with seven polymorphic CFTR DNA markers. *Hum Mutat* 1996: 8: 149–159.

Noone PG, Knowles MR. "CFTR-opathies": disease phenotypes associated with cystic fibrosis transmembrane regulator gene mutations. *Respir Res* 2001: 2: 328-332.

Palomaki GE, FitzSimmons SC, Haddow JE. Clinical sensitivity of prenatal screening for cystic fibrosis via CFTR carrier testing in a United States panethnic population. *Genet Med* 2004: 6: 405-414.

Radivojevic D, Djurisic M, Lalic T et al. Spectrum of cystic fibrosis mutations in Serbia and Montenegro and strategy for prenatal diagnosis. *Genet Test* 2004: 8:276-80.

Riordan JR, Rommens JM, Kerem B et al. Identification of the cystic fibrosis gene: cloning and characterization of complemantary DNA. *Science* 1989: 245: 1066–1073.

Rommens JM, Iannuzzi MC, Kerem B et al. Identification of the cystic fibrosis gene. chromosome walking and jumping. *Science* 1989: 245: 1059–1065.

Sepuvelda W, Sebine NJ. Fetal echogenic bowel: a complex scenario. *Ultrasound Obstet Gynecol* 2000: 16: 510–514.

Strocker AM, Snijders RJ, Carlson DE et al. Fetal echogenic bowel: parameters to be considered in differential diagnosis. *Ultrasound Obstet Gynecol* 2000: 16: 519–523.

Strong TV, Wilkinson DJ, Mansoura MK et al. Expression of an abundant alternatively spliced form of the cystic fibrosis transmembrane conductance regulator (CFTR) gene is not associated with a cAMP-activated chloride conductance. *Hum Mol Genet* 1993: 2: 225-230.

Super M, Schwarz MJ, Malone G, Roberts T, Howorth A, Dermody G. Active cascade testing for carriers of cystic fibrosis gene. *Br Med J* 1994: 308: 1462–1468.

Trezise AEO, Linder CC, Grieger D et al. CFTR expression is regulated during both the cycle of semineferous epithelium and the oestrous cycle of redents. *Nat Genet* 1993: 3: 157-164.

Tzetis M, Kanavakis E, Antoniadi Th, Doudounakis S, Adam G, Kattamis C. Characterization of more than 85% of cystic fibrosis alleles in the Greek population, including five novel mutations. *Hum Genet* 1997: 99: 121–125.

Tzetis M, Efthymiadou A, Strofalis S et al. CFTR gene mutations- including three novel nucleotide substitutions- and haplotype background in patients with asthma, disseminated bronchiectasis and chronic obstructive pulmonary disease. *Hum Genet* 2001a: 108: 216-211.

Tzetis M, Efthymiadou A, Doudounakis S, Kanavakis E. Qualitative and quantitative analysis of mRNA associated with four putative splicing mutations (621+3A>G, 2751+2T>A, 296+1G>C, 1717–9T>C-D565G) and one nonsense mutation (E822X) in the CFTR gene. *Hum Genet* 2001b: 109: 592–601.

Van der Ven K, Tsui LC, Boat TF et al. Cystic Fibrosis mutation screening in healthy men with reduced sperm quality. *Hum Reprod* 1996: 11: 513-517.

Vankeerbergerghen A, Cuppens H, Cassiman JJ. The Cystic Fibrosis transmembrane conductance regulator: an intriguing protein with pleiotropic functions. *J Cystic Fibrosis* 2002: 1: 13-29.

Vrettou C, Tzetis M, Traeger-Synodinos J, Palmer G, kanavakis E. Multiplex sequence variation detection throughout the CFTR gene appropriate for preimplantation genetic diagnosis in populations with heterogeneity of cystic fibrosis mutations. *Mol Hum Reprod* 2002: 8: 880-886.

Welsh MJ, Ramsey BW, Accurso F, Cutting GR. Cystic Fibrosis. In: Sly WS, Valle D, eds. The Metabolic and Molecular Bases of Inherited Disease, 8th edn, Vol. 3. New York: McGraw-Hill 2001, pp. 5121–5188.

Zielenski J, Rozmahel R, Bozon D et al. Genomic DNA sequence of the cystic fibrosis transmembrane conductance regulator (CFTR) gene. *Genomics* 1991: 10: 214-228.

Zielenski J, Tsui LC. Cystic Fibrosis: genotypic and phenotypic variations. *Annu Rev Genet* 1995: 29: 777–807.

Zielenski J. Genotype and Phenotype in cystic fibrosis. *Respiration* 2000: 67: 117–133.

In: Focus on Genetic Screening Research
Editor: Sandra R. Pupecki, pp. 27-55
ISBN 1-60021-229-8

Chapter II

Anxiety of Pregnant Women after Prenatal Amniocentesis and Genetic Testing

Fusun Terzioglu*
Hacettepe Universities; Hemşirelik Yüksekokulu;
06100-Ankara/TURKİYE

Abstract

Background

Prenatal genetic screening tests are used to analyze the number of chromosomes and to detect the chromosomal abnormalities such as Down syndrome during the early period of pregnancy. Genetic tests enable couples to be informed beforehand about the health of their child and offer the opportunity to terminate an affected pregnancy. However, genetic screening testing following an amniocentesis procedure in the prenatal period, and a positive test result cause anxiety in pregnant women.

Aim

This study is carried out as descriptive to evaluate the distress of pregnant women in relation to genetic testing, their anxiety level and the effects of risk factors on anxiety.

* Correspondence Address: Associate.Prof. Fusun Terzioglu; Hacettepe Universities; Hemşirelik Yüksekokulu; 06100-Ankara/TURKİYE; Tel [Mobil]:+90 532 4400004; Tel [Office]: + 90 312 3242013; + 90 312 3051580/121; Fax: + 90 312 3127085; E-Mail: fusun@hacettepe.edu.tr ; fusun_terzioglu@yahoo.com

Method

In this study 130 pregnant women who had undergone genetic testing following an amniocentesis were interviewed. A questionnaire was used for data collection. Anxiety of the women was assessed using the Spielberger's state-trait anxiety inventory (STAI). Completing the questionnaire form and anxiety scales lasted about half an hour.

Results

The majority of the women who underwent amniocentesis expressed distress prior to genetic testing. The mean trait anxiety scores of women before receiving the test results was 43.72 and the state anxiety scores of women after the test result was 36.05. The association of state anxiety level of pregnant women after they received the test results indicated, those women who had no living child, who had a child with abnormality, who reported one or more spontaneously abortion history, who were younger, and who have a positive test result revealed significantly higher anxiety level. In contrast, there was no significant association between the trait anxiety of women and the education level, having a consanguine marriage, and the need for counseling following the test results.

Conclusion

In this study it was identified that women referred for genetic testing after an amniocentesis procedure experienced distress before testing, they were anxious and psychologically affected by undergoing amniocentesis and genetic testing, even if the test results were negative. Perceiving not to be informed in detail about genetic testing and amniocentesis raised their anxiety level. The results revealed the importance of a comprehensive genetic counseling for women who underwent amniocentesis and genetic testing.

Key Words: Prenatal genetic testing; psychological distress; amniocentesis; anxiety; nursing role.

Introduction

Developments in biomedical technology and the information obtained from the Human Genome Project are offering unforeseen opportunities for predicting the future health of individuals and determining the genetic basis of the population. These developments have resulted with new genetic tests, and knowledge in the field of genetics that will open new horizons for reproductive health.

Genetic tests are diagnostic techniques used to analyze a specific gene, its product or function that is likely to be associated with a specific genetic disorder or mutation. Genetic tests are used to provide a more precise diagnosis for a suspected genetic disorder, to identify a carrier status or to detect the predisposition to certain genetic diseases and conditions. Genetic tests are also used for prenatal diagnosis to determine the number of chromosomes of fetus during the intrauterine life, and to detect the chromosomal abnormalities such as Down syndrome or Turner syndrome [Williams and Lessick, 1996; Raines, 1998; Collins, 1999; Harper et al., 2001].

There are three common types of genetic tests: cytogenetic tests, biochemical analyses and molecular tests. Cytogenetic tests involve the evaluation of human chromosomes and their abnormalities. Chromosomal analysis for prenatal diagnosis usually requires obtaining fetal cells from the amniotic fluid. Biochemical tests are standard laboratory tests used to determine the congenital metabolic diseases, or conditions associated with the lack of specific metabolic enzymes. For example, phenylketonuria is detected by biochemical tests. Molecular tests include the analyzing of DNA alterations to diagnose a certain genetic disease more precisely [Tinkle and Cheek, 2002].

Genetic testing enables individuals to be informed about their health status and risks, help them in making accurate decisions for their health and offer the opportunity for early diagnosis and treatment. However, depending on the aim and result, genetic testing can also cause psychosocial, ethical and legal problems. Debates surrounding genetic testing often focus on the misuse of genetic information. Potential harms of genetic testing include stigmatization, discrimination, violation of privacy and confidentiality, alteration in self-esteem, disruption of parents and child relationships, and anxiety and depression [Juengst, 1994; Haris and Haris, 1995; Roth and Painter, 2000; Jeffords and Daschle, 2001; Vastag, 2001; Burke and Pinsky, 2001; Paabo, 2002]. In particular, the pre-implantation genetic tests used in prenatal screening have a potential to make embryo selection possible because they involve the transfer of healthy embryos into the uterus and the destruction of many embryos. In the long term, these tests can be used not only for the selection of healthy embryos, but also for eugenic aims including the selection of embryos on the basis of gender, mental status, skin color, etc., which can lead to a black market for embryo selection [Raines, 1998; King, 1999; Draper and Chadwick, 1999; Harper et al., 2001; Lavery et al., 2002; Giarelli, 2003].

Prenatal screening tests, the most well-known among the genetic tests, are used to detect the number and alteration of chromosomes of the fetus [Raines, 1998]. Prenatal diagnostic tests are acknowledged as the most important developments in the field of genetics. Prenatal diagnosis includes all phases of embryonic and fetal screening and about 8% of all pregnancies have an indication for prenatal diagnosis [Saraçoglu, 1998].

Prenatal screening tests include various genetic tests used to detect and/or analyze the reasons for infertility, miscarriages, stillbirths, neonatal death, multiple malformations, developmental and mental retardations, psychological disorders, etc. Finley [1982] points out that genetic factor are responsible for one out of 5 infant deaths, one out of 4 mental retardations, and almost half of the IQ level below 50. Literature indicates that every 4 baby in every 100 living births (%4) may have an abnormality, which most of the reasons of such abnormalities are not known. Studies report that 0.6% of the new birth anomalies arise from a chromosomal disorder, 1% from a single gene defect, which cause about 5000 hereditary diseases, and at least 2% arise out of structural malformations. Spina bifida, malformations of the heart and cleft palate are among these structural malformations [O'Brien, 1984; Schemmer and Jhonson, 1993; Harper et al., 2001].

Prenatal screenings for genetic purposes are most effective in detecting the carrier status of couples for recessive genetic disorders such as Tay Sach's disease, Beta Thalassemia and sickle cell anemia [Gitzelman, 82]. Recessive inherited genetic disorders are more problematic in areas where consanguinity marriages are common. For example in Turkey, the rate of intermarriage among blood relatives is about 25.1% [Hancıoglu et al., 2001]. This may increase the risk of fetus with congenital anomalies and autosomal recessive inherited diseases, suggesting the importance of genetic counseling for individuals contemplating marriage with their relatives.

In contrast, chromosomal abnormalities, such as Down syndrome may not be inherited from parents. There may also be birth defects in the fetus whose genetic components are not the sole cause, but occur due to the exposure to some agents such as drugs, radiation, etc., which cause mutation in genes [Gardner and Sutherland, 1989; Kolker, 1989; Connor and Ferfuson, 1993].

Today, diagnostic ultrasound and triple screening test are in widespread usage to detect the above-mentioned pathological conditions. Although these diagnostic methods offer valuable information about the development of fetus and its anomalies, they are limited in providing detailed information [Quilligan and Zuspan, 2000]. In this case, a higher and invasive prenatal diagnostic procedure, such as genetic amniocentesis (GA) and chorionic villus sampling (CVS) may be required [Sjogren and Uddenberg, 1989; Hunfeld et al., 1996; Quilligan and Zuspan, 2000; Grant, 2000; Harper et al., 2001].

With the use of CVS and genetic amniocentesis it is possible to identify all of the chromosomal disorders, some of the single gene disorders and structural malformations during the early period of pregnancy [O'Brien, 1984; Schemmer and Jhonson, 1993; Harper et al., 2001; Lavery et al., 2002]. Among these methods, genetic amniocentesis is used as the first and most common diagnostic procedure in identifying fetal disorders. Genetic amniocentesis has been used since the beginning of the 1970s and offers the opportunity to diagnose the specific conditions of fetus [Haddow et al., 1998; Robinson et al., 1998; Browner et al., 1999].

Amniocentesis is a procedure that includes the analyzing of cells obtained from the amniotic fluid around the 14-16th week of gestation. Amniocentesis is usually performed following a local aesthesia of the abdominal skin, using a 20 or 22 gauge, 7-15 cm long needle to reach through the mother's abdomen into the amniotic sac with the guidance of ultrasound. Approximately 30 cc of amniotic fluid is aspirated. Fetal cells that spill out from the skin, connected tissue, respiratory, gastrointestinal and urinary system of the fetus and amniotic membrane are isolated to the flash medium, which contains antibiotic fluids. Culture of the cells takes around 10-15 days, rarely 3 to 4 weeks. If sufficient cells and culture were obtained, the sample is transferred from the incubator for preparation procedure. Usually 15 cells are used for analyzing. Chromosomes are identified and counted one by one by staining technique, which gives a characteristic pattern of alternating light and dark bands. Photos of chromosomes of one or more cells are taken. Each of the chromosomes are matched with its pair and placed according to its numeric sequence. This procedure is defined as karyotyping and performed very easily and quickly through the use of high visual analyzing systems. However, in 1% of the cases, there may be a need to repeat the test. The reasons for re-testing usually include obtaining insufficient amniotic fluid or problems related to laboratory procedures. Sometimes, because of unknown reasons fetal cells can not be cultured and analyzed. Excessive bloody sample and bacterial contamination might cause failures in the culture of cells [Tabor et al., 1986; Pillitteri, 1992; Kowacek et al., 2002].

The prenatal diagnostic procedure, amniocentesis, is a useful method for the following indications: [O'Brien, 1984; Tabor et al., 1986; Gardner, 1989; Pillitteri, 1992; Schemmer and Jhonson, 1993; Grant, 2000; Kowacek et al., 2002].

- Advanced maternal age >35
- Chromosomal abnormality in one or both of the couples
- History of three or more miscarriages during the previous pregnancies of the mother
- Abnormal ultrasound scan that provoke suspicion about the chromosomal abnormalities of the fetus
- Abnormal levels of biochemical markers in the maternal blood [alpha fetoprotein, beta human chorionic gonadotrophin, unconjugated estriol and triple screen test]
- Positive family history of genetic disorders (e.g. single gene defects such as muscular dystrophy, hemophilia, thalassaemia and those couples referred for DNA analyzing)
- A psychological indication in 10-20% of cases due to the concern of couples about the health of their baby [Sjogren, 1996]
- Previous chromosomal abnormality or mental retardation in child of unknown etiology
- Drug use or long-term exposure to teratogens or mutagens
- Pregnancies after treatment for infertility (intracytoplasmatic sperm injection)

The main indication for amniocentesis in clinical usage is advanced maternal age. However, because of various indications, the use of genetic amniocentesis in women below the age of 35 years has increased rapidly. For example, the rate of Down syndrome due to invasive procedures was around 16% in women over the age of 40, and 35% in women over 35 years. While 25-35% of all Down syndrome cases occurs in pregnancies over the age of 35, 70-80% of Down syndromes are identified in younger pregnancies [Kolker and Burke, 1994; Ferber et al., 2002]. Therefore, although the advanced maternal age is the main indication for genetic amniocentesis; pregnant women in all age groups should be taken into consideration.

In this respect, genetic amniocentesis is a valuable diagnostic method in identifying the risk factors for fetal abnormalities during the prenatal period. However, there are several risks associated with amniocentesis, including the fetal injury, rupture of amniotic membrane, maternal infection, fetal and maternal hemorrhage and the risk of miscarriage. The risk of infection is around 1:200 to 1:400. Pregnancy loss rates following amniocentesis are around 1:200 to 1:300 [Kolker, 1989; Chescheir and Hansen, 1999; Teryack et al., 2001]. When compared to other invasive procedures, most of the patients perceived it as stressful and painful [Beeson and Golbus, 1979; Tabor et al, 1986; Tabor and Jonsson, 1987; Evers-Kiebooms et al., 1988; Marteau et al., 1988; Black, 1989; Robinson et al., 1989; Weinans et al., 2000; Fischer et al., 2000; Tercyack et al., 2001; Haris et al., 2004]. All of these potential risks may raise the anxiety of couples, especially pregnant women before genetic testing.

Apparently, genetic amniocentesis has advantages and disadvantages, as all the other prenatal screening methods that affect the couples physically, psychologically and socially. Genetic amniocentesis affects pregnant women physically, because it is an invasive procedure that causes pain and requires local anesthesia. Pregnant women are also affected psychologically due to the feelings of guilt and regret about the potential of procedure-related fetal injury. They may also be concerned about the underlying fetal abnormalities and the risk of loss. Research suggests that prenatal diagnostic methods can be anxiety provoking due to various concerns of individuals about the effects of these methods [Astbury and Walters,

1979; Jorgensen, 1987; Hunfeld et al., 1993; Hunfeld et al., 1996; Evers-Kiebooms et al., 1998; Cederlhom et al., 1999; Fischer et al., 2000].

In addition, studies also demonstrated that genetic amniocentesis and a positive test result, after undergoing genetic testing, causes a considerable level of anxiety. Abuelo et al. [1991] found in their study, higher levels of anxiety in women with low alpha-fetoprotein. Evers-Kiebooms et al. [1998] investigated the psychological effect of amniocentesis in three different groups [women in advanced age, mothers who had a previous child with Down syndrome and mothers who had a previous child with neural tube defect]. They found different anxiety levels between and within groups. Tercyak et al. [2001] demonstrated that the anxiety of pregnant women was clinically elevated and highest at the pre-counseling stage and that previous experience with prenatal diagnostic testing was associated with increased pre-counseling anxiety. Feeny et al. [2002] assessed the effect of chorionic villus sampling and genetic amniocentesis on the perception of health-related life quality of individuals and they found significant differences between the anxiety and the perception of health-related life quality. Fischer et al. [2000] conducted a study to determine the effect of light pressure effleurage on both perception of pain and overall anxiety associated with genetic amniocentesis. They concluded that amniocentesis is an anxiety provoking procedure. Robinson et al. [1998] reported an increased anxiety level of pregnant women before the amniocentesis procedure. Philips and Zinn [1986] and Van Zuuren [1993] investigated the relationship between amniocentesis and coping mechanisms and reported that women experience intense psychological distress and elevated anxiety before and during the amniocentesis procedure. Ng et al. [2004] pointed out that merely being aware of undergo genetic amniocentesis causes anxiety in pregnant women. Lavery et al. [2002] assessed the experiences and attitudes of patients who received genetic counseling, and they reported that 41% of patients found the treatment process extremely stressful.

All of these studies underscore the importance of an effective genetic counseling before prenatal genetic screening and after genetic testing. Many studies have indicated that an effective genetic counseling reduces anxiety [Robinson et al., 1989; Kolker, 1989; Abuelo et al., 1991; Ormond et al., 1996; Michie et al., 1997; Evers-Kiebooms et al., 1998; Ng et al., 2004; Grant, 2000; Tercyack et al., 2001; Fenny et al., 2002]. The increasing incorporation of new genetic tests into clinical practice, and the demand for knowledgeable and skilled health professionals in the field of genetics to provide information and counseling for individuals reveal the importance of this subject. Nurses can help individuals in understanding the benefits and potential risks/ limitations of genetic screening tests by providing them with the relevant information. They can also provide support for the pregnant women during the antenatal period by informing them timely about the resources.

Nurses can play a key role by identifying risk factors, evaluating individuals and their families, assessing family history, providing genetic information and counseling, keeping records, and referring individuals to genetic experts and services [Wertz et al., 1994; Raines, 1998; Jenkins, 2000; Middelton et al., 2002; Tinkle and Check, 2002; Giarelli, 2003; Terzioğlu and Dinç, 2004]. Furthermore, the expanding role of nurses in the field of genetics requires them to evaluate the problems and needs of individuals. Nursing is a profession that is centered on human needs. Therefore, the potential impact of genetic testing on couples, pregnant women and families should be considered in planning the necessary nursing care.

Purpose of the Research

The purpose of this descriptive research is to investigate the effects of genetic testing on pregnant women who underwent amniocentesis in the prenatal period. This study can contribute to the research on this subject, provide assistance to the nurses and midwifes who deliver care, and provide education and counseling to the individuals in prenatal care units by increasing their awareness of the potential physical and psychological influences of prenatal screening tests, in particular, the effects of genetic amniocentesis.

The Research Questions Addressed in this Study Were

1. Do pregnant women who are referred for genetic testing after amniocentesis experience psychological distress and anxiety prior to genetic test results?
2. Does the level of anxiety vary by the test results and is there a difference in levels of anxiety between women who received a positive test result and negative test result?
3. Is there a difference in the levels of state anxiety and trait anxiety on pregnant women?
4. Is there a difference in the levels of state anxiety between young and advanced maternal age?
5. Does the level of anxiety vary by having a previous child with an abnormality?
6. Does the level of anxiety vary by having a consanguineous marriage?
7. Does the level of anxiety vary by the need of genetic counseling?

2. Methodology

Study Site and Sample

For the aim of this study, two university hospitals were selected as the study site, because the state hospitals and private hospitals within the municipality of Ankara have no genetic departments and do not perform genetic testing. Both university hospitals offer genetic testing in their genetic departments and practice obtaining informed consent routinely through a printed consent form used by physicians who are experts on genetics.

The sample size consisted of 130 pregnant women referred for chromosomal analysis after amniocentesis from obstetric units of hospitals between January and June, 2003. This study only considered pregnant women who were undergoing genetic testing for the first time, who had received the test results and who agreed to participate in this study.

Ethical Considerations

Official and ethical approval from the relevant institutions for this study was obtained. The pregnant women coming to genetic departments to receive the test results were informed about the aim and method of the study; they were told that their participation was voluntary, and that they had the right to withdraw at any point.

Data Collection

The author of this study collected data through using a semi-structured interview method. Semi-structured interviews were chosen, because it would be difficult to explore the psychological response of genetic testing on pregnant women, especially the feelings upon receiving the test results with closed-ended questions or by using the anxiety inventory alone, due to the differences in cultural backgrounds of participants that come to the genetic departments from various geographical areas of the country. Semi-structured interviews involved the preparation of a data collection form that served as an interview guide. The data collection form included multiple-choice questions on demographic characteristics of individuals and several open-ended questions to examine the influence of genetic testing, the support and counseling needs of the participants. Open-ended questions included the following:

- 'Did you experience distress before genetic testing?'
- 'How did you feel when you received the test results?
- 'Do you think that you will need psychological support or genetic counseling following the test results?'

Anxiety level of the women was measured by using the Spielberger's State and Trait Anxiety Inventory (STAI). STAI is a 40-item scale that is used to measure trait and state anxiety with an internal consistency reliability ranging from alpha = 0.94 and 0.96. It was adapted to Turkish, and validated between 1974-1977 by Öner and Le Compte [1982]. An investigator completed the data collection form during the interview and then the women were asked to fill out the trait anxiety inventory. When pregnant women received the genetic test results, they were asked to fill out the state anxiety form. Interviewing and filling the STAI inventory took approximately 30 minutes.

Data Evaluation

The data were evaluated with SPSS 10.0 (Statistical Package for Social Sciences for Windows) packet program using percentage, Pearson and Fisher's exact chi square. As the aim was not to make a comparison between hospitals, the data obtained from pregnant women of the two hospitals were evaluated together. Demographical data and responses to closed-ended questions were evaluated by percentages. Similar types of responses of participants to the open-ended questions were first grouped into single categories then a code number was assigned to each group and entered into the computer. Anxiety scores were calculated in line with the instructions of Spielberger, and mean scores were obtained. In order to detect the normal distributions of the data, the Kolmogorov-Smirnov test was used. Since the data was normally distributed, we used the independent samples *t* test and ANOVA for comparing the mean anxiety scores between groups. Statistical significance was defined as $p<0.05$.

3. Results

Socio-Demographic Characteristics of Pregnant Women

The results of this study, which was carried out to investigate the effect of genetic testing on pregnant women who underwent amniocentesis in prenatal period are the following:

The mean age of pregnant women was 32.91 ± 5.7 years. About 54.6% of the women were graduates from a university or had a higher education level. More than half of the women had a professional career (61.5%) and reported their economical status as moderate (65.4%). According to the research of TNSA [2003] in Turkey, 42% of all women were working and about 17% had an education level of high school or higher. The reason of higher education level of the participants in this study may be related to the sample, which was selected from the university hospitals of the capital city, where the education level of the population is relatively higher. The majority of the pregnant women were living in an urban area (81.5%) and the vast majority (96.2%) had health insurance (Table 1).

Table 1. Socio-Demographic Characteristics of Pregnant Women

Characteristics	Frequency N=130	Percentage
Age		
18-26	21	16.1
27-34	46	35.4
35 and up	63	48.5
Mean Age	32.91±5.72; (min=20. max=42)	
Education Level		
Primary School	25	19.2
Secondary School	34	26.2
University	71	54.6
Profession		
Yes	80	61.5
No	50	38.5
Economical Status		
Good	36	27.7
Moderate	85	65.4
Bad	9	6.9
Residence		
Urban area	106	81.5
Rural area	24	18.5
Health Insurance		
Yes	125	96.2
No	5	3.8
Total	**130**	**100.0**

The mean marriage age of women who underwent amniocentesis and genetic testing was 24.88±5.78 (min=14; max=39). In Turkey, marriage is culturally supported and only 29.9% of women never marry during their lifetime. Although the average marriage age of women varies according to the regions, the median age is reported as 22 [TNSA, 2003]. The reason for the higher marriage age of the participants in this study may be related to the finding that most of the pregnant women were living in an urban area and having a higher education level, which affects their decision on delaying the marriage age.

Obstetric and Family History of Pregnant Women

The average age of first pregnancy in our participants was 26.34±5,74 (min=15; max=39). The average age of first pregnancy in Turkey was reported as 22 years [TNSA, 2003]. Twenty percent of the women reported that they are primigravida, 57.7% had one or more living child, 35.4% had a previous experience of one or more spontaneous abortion and 25.4% had selected abortion (Table 2). In Turkey, the rate of spontaneous abortion is 10% and selected abortion is 11.3% [TNSA, 2003]. The reason for higher rates of spontaneous abortion in this study may be due to the risk factors associated with the current pregnancy of the women, such as their advanced age, consanguinity marriage and positive history of abortion.

The indications for amniocentesis in 97.8% of women referred to the genetic department were: advanced maternal age, consanguinity marriage, a history of miscarriage, stillbirth or neonatal birth, abortion, having a previous child with abnormalities, and higher or lower level of alpha fetoprotein.

Consanguinity Marriage, Positive History of a Baby with Anomalies, and Results of the Genetic Test

About 20.8% of pregnant women were married with their blood relatives, among those 59.3% had a first-degree consanguinity marriage. Approximately 17% of pregnant women had a previous child with physical or mental abnormality (Table 3). It was identified that 93.1% of women who underwent genetic testing after amniocentesis received a normal test result, and 6.9% a positive test result (Table 4). Of those with abnormal test results, 55.6% had chromosome abnormality, 22.2% single gene defect, 22.2% multi-factorial disorders. The positive test results indicated the chromosomal abnormalities, including Trisomy 21/13/ mosaic (one person), 47 XX (two person), 46 XY t(6:19) (q22:q32) (one person), 45, X turner syndrome (one person), 46 XX t(8:18) (one person); 45 XY t(4:21) (one person), Thalassemia minor/major (one person), leukodystrophy (one person) and single gene defects and multifactorial defects. There was a significant association between the positive test results of pregnant women and having a previous child with anomalies (p=0,001) (Table 5). It was identified that 40.7% of women who had a previous child with an abnormality also had a consanguinity marriage (p=0,001) (Table 6). In contrast, there was no significant association with regard to the positive genetic test result and consanguinity marriage, among those who received a positive test result, only 14.8% had a consanguinity marriage (p>0,05) (Table 5).

Table 2. Obstetric History of Pregnant Women

Obstetric Characteristics	Frequency	Percentage
Age of First Pregnancy		
18 and below	10	7.7
19-29	86	66.2
30-39	34	26.1
Mean Age of First Pregnancy	X=26.34±5.74; (min=15. max=39)	
Number of Pregnancy		
Primigravida	26	20
2 and up	104	80
Para		
None	46	35.4
1-2	78	60.0
3 and up	6	4.6
Living Child		
None	55	42.3
1-2	72	55.4
3 and up	3	2.3
Spontaneous Abortion		
None	84	64.6
1-2	32	24.6
3 and up	14	10.8
Selected Abortion		
None	97	74.6
1-2	29	22.3
3 and up	4	3.1
Total	**130**	**100.0**

Table 3. Status of Consanguinity Marriage and Having a Previous Child with Abnormality of Pregnant Women

Variables	Frequency	Percentage
Consanquinity Marriage		
Yes	27	20.8
No	103	79.2
Relative Degree		
First Degree	16	59.3
Second Degree	11	40.7
Previous Child with Abnormality		
Yes	22	16.9
No	108	83.1
Total	**130**	**100.0**

Table 4. Experienced Distress Before Genetic Testing . Results of The Genetic Test and Counseling Need About Test Results of Pregnant Women

Some Variables	Frequency	Percentage
Distress Before the Genetic Testing		
Yes	126	96.9
No	4	3.1
Results of the Genetic Test		
Normal	121	93.1
Abnormal	9	6.9
Counseling Need About Test Results		
Yes		
No	80	61.5
	50	38.5
Total	**130**	**100.0**

Table 5. Relation Between Consanquinity Marriages and Status of Having Previous Child With Abnormality of Pregnant Women Who Were Subjected to Genetic Tests

Results of the Test	Consanquinity Marriages		Previous Child with Abnormality	
	Yes	No	Yes	No
Normal	23 (85.2)	98 (95.1)	15 (68.2)	106 (98.1)
Abnormal	4 (14.8)	5 (4.9)	7 (31.8)	2 (1.9)
Statistical Analysis	X^2=3.294. p=0.089		X^2=25.470. **p=0.001**	

Table 6. The Relation between Consanquinity Marriage and Having Previous Child with Abnormality According to Genetic Tests

Consanquinity Marriage	Previous Child with Abnormality	
	Yes	No
Yes	11 (40.7)	16 (59.3)
No	11 (10.7)	92 (89.3)
Statistical Analysis	X^2=13.750. **p=0,001**	

Being Informed Prior to Genetic Testing and Distress

A vast majority of pregnant women (96.9%) reported psychological distress before undergoing genetic testing (Table 4). When they were asked what kind of a distress they experienced, most of them expressed the concerns related with the risks of amniocentesis. These concerns include the risk of abortion and fear of bleeding, the risk of infection or the potential of fetal injury. Since most of the women in this study were well educated, which affects their knowledge on the risks of amniocentesis, and most had undergone amniocentesis just before genetic testing, their concerns are understandable.

When the information status of women has been questioned, 26.2% of the women stated that they were not informed about the aim, 30.8% about procedure, 46.2% about the potential risks, and 9.2% about the results of the test (Table 7). The rate of women who reported to be informed completely about the aim, procedure, risks and results of genetic testing was one-third. The remaining stated that they were partially informed. Seventy-two percent of women were informed completely about the test result (Table 7), but expressed that they did not understand the reason for genetic testing or the problem they had to face.

Table 7. Effects of Test Procedure and Level of Knowledge about The Genetic Test of Pregnant Women after Amniocentesis

Some Variables	Frequency	Percentage
Effect of Test Procedure		
Yes	95	73,1
No	35	26,9
Information About Aim of the Test		
Yes	48	36,9
Partially	48	36,9
No	34	26,2
Information About Procedure of the Test		
Yes	45	34,6
Partially	45	34,6
No	40	30,8
Information About Risks of the Test		
Yes	41	31,5
Partially	29	22,3
No	60	46,2
Information About the Results of the Test		
Yes	94	72,3
Partially	24	18,5
No	12	9,2
Total	**130**	**100,0**

Anxiety Scores and Associations of Scores with Independent Variables

The trait anxiety scores of pregnant women was 43,72± 8.48, and state anxiety scores was 36,05±12,25. Differences between trait and state anxiety scores were found statistically significant (p=0,001) (Figure 1). Öner and Le Compte, reported that the moderate anxiety level of healthy individuals ranged from 36 to 42. Compared to their report, the trait anxiety level of pregnant women in this study is higher.

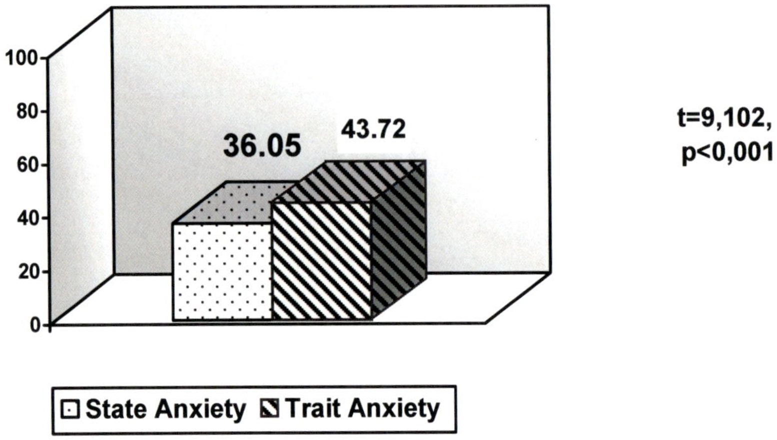

Figure 1. State and Trait Anxiety Mean Scores of Pregnant Women

The state anxiety scores of women who reported to be informed about the aim, procedure, and risks of genetic testing were significantly elevated than those who claimed not to be informed (Figure 2; Table 9). In contrast, the state anxiety scores of women who reported that they consented for genetic testing were significantly higher than those who stated not being asked for consent (p=0,015).

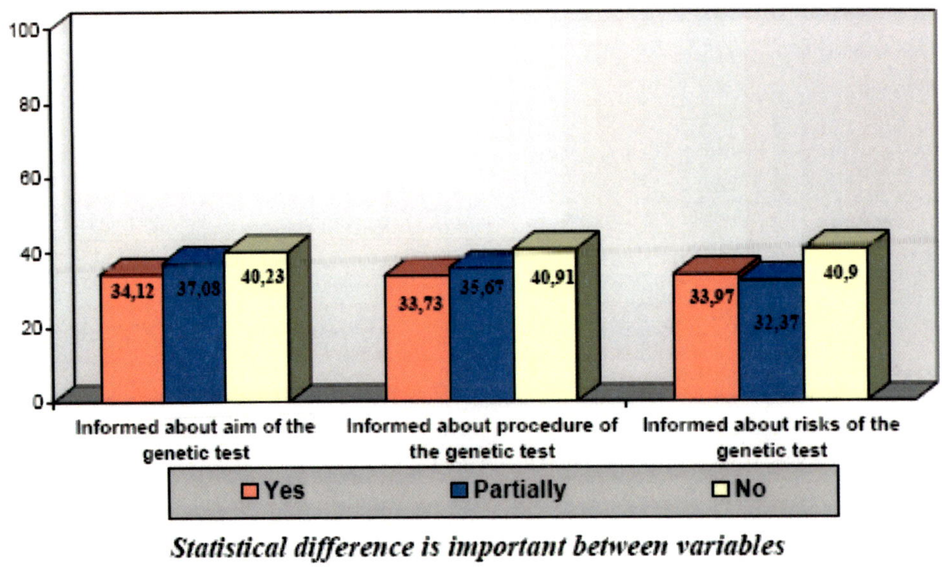

Figure 2. State Anxiety Mean Scores of Pregnant Women After Amniocentesis According to their Level of Knowledge About Test Aim and Process and Risks

There was significant association between the anxiety scores of women, and test results, being childless and previous experience of abortion. Women who received a positive test result had higher anxiety scores (53,33±10,4) than those who received a negative test result (35,58±11,4) (p=0,001] (Figure 3). The state anxiety scores of childless women (39,21±13,2) were also higher than those who had one or more living child (35,05±11,2) (p=0,05). Likewise, the state anxiety scores of women with previous experience of one or more spontaneous abortion were higher (42,13±12,4) than those who never had an abortion (34,01±11,8) (p=0,001) (Table 8).

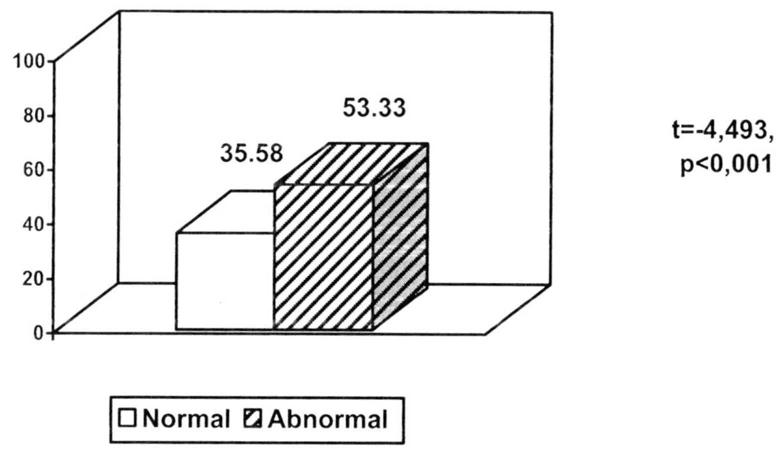

Figure 3. State Anxiety Mean Scores of Pregnant Women in Relation to the Genetic Test Results

In addition, the state anxiety scores of women who had a previous baby with anomaly during their past pregnancies were higher (42,59±13,4) than those who did not have a baby with anomalies (35,63±11,7) (p=0,015) (Table 8). With regard to the association between the test results and having a previous baby with anomaly and consanguinity marriage, it was found that the state anxiety scores of those who did not have a consanguinity marriage and a previous baby with anomaly were higher (p=0,001).

With regard to the association between the state anxiety scores of women and age groups, it was found that young women scored higher (41.38±13,2) than those in advanced age (34,09±10,7) (p=0,038) (Figure 4).

Women who had a consanguinity marriage also had higher elevated state anxiety scores (38,42±14,0) than those who did not (35,43±11,2). However, contrary to expectations, there was no statistically significant association between these variables (p>0.05) (Table 8).

The level of education, age of pregnancy, frequency of pregnancy, voluntary abortion history, the need of counseling according to the test results, employment status, residence region, and having health insurance were not associated with anxiety scores of pregnant women (Figure 5).

Table 8. State and Trait Anxiety Mean Scores of Pregnant Women in Relation To Obstetric History and Consanguinity Marriage

Independent Variables	N	State Anxiety	Trait Anxiety
Consanguinity Marriage			
Yes	27	38.42±14.03	47.44±7.6
No	103	35.43±11.22	42.74±8.00
Statistical Analysis		t=-0.357. p=0.722	t=-4.827. **p=0.001**
Level of Relatives			
First degree	16	40.18±14.24	49.50±6.20
Second degree	11	36.54±16.69	47.36±9.36
Statistical Analysis		t=-0.609. p=0.548	t=0.715. p=0.481
Number of Pregnancy			
Primigravida	26	37.57±13.78	44.69±8.79
2 and up	104	36.62±11.90	44.52±8.23
Statistical Analysis		t=0.353. p=0.725	t=0.089. p=0.929
Number of Child			
None	55	39.21±13.20	46.85±8.17
1 and up	75	35.05±11.26	42.88±8.06
Statistical Analysis		t=1.935. p=0.055	t=2.760. **p=0.007**
Spontaneous Abortion			
None	84	34.01±11.08	43.08±8.09
1 and up	46	42.13±12.48	46.93±8.25
Statistical Analysis		t=-3.745. **p=0.001**	t=-2.515. **p=0.013**
Previous Child with Abnormality			
Yes	22	42.59±13.40	46.90±8.87
No	108	35.63±11.72	44.08±8.15
Statistical Analysis		t= 2.473. **p=0.015**	t=1.459. p=0.147

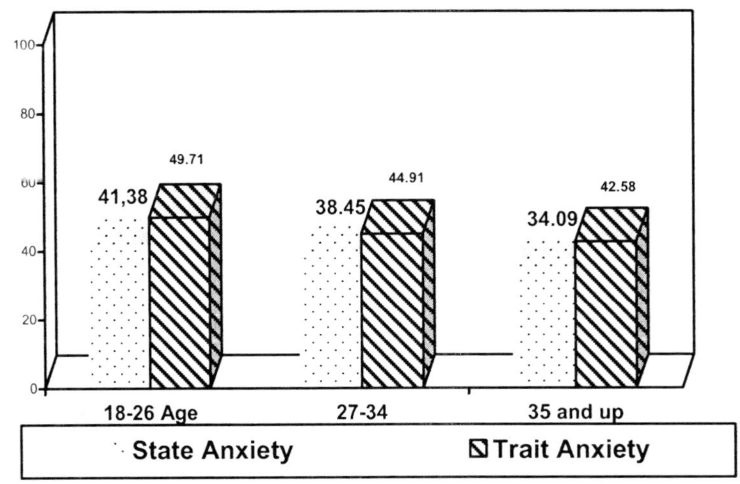

Statistical difference is important between variables

Figure 4. State and Trait Anxiety Scores of Pregnant Women After Amniocentesis According to Ages.

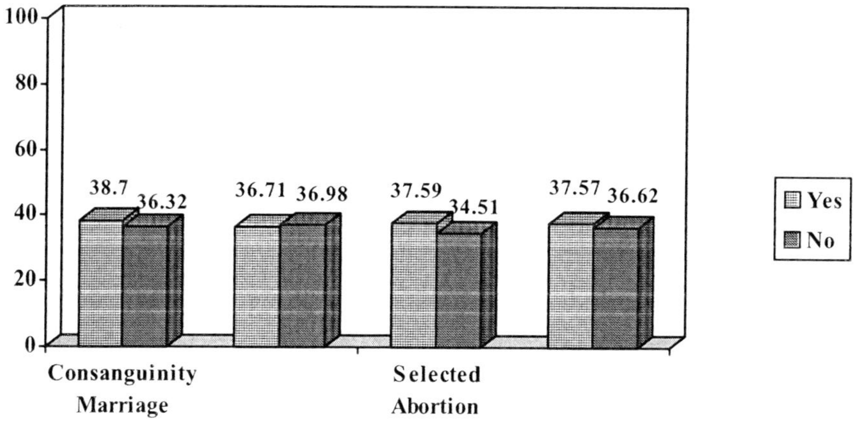

No statistical difference between groups p>0,05

Figure 5. State Anxiety Mean Scores of Pregnant Women According to Research Criteria

Table 9. State Anxiety Mean Scores of Pregnant Women According to Level of Knowledge about Genetic Tests after Amniocentesis

Variables	N	State Anxiety
Informed about the Aim of the Test		
Yes		
Partially	48	34.12±10.84
No	48	37.08±13.53
	34	40.23±11.64
Statistical Analysis		F= 2.553. p=0.08
Informed about the Procedure of the Test		
Yes	45	33.73±10.68
Partially	40	35.67±13.77
No	45	40.91±11.38
Statistical Analysis		F=4.323. **p=0.015**
Informed about the Risks of the Test		
Yes	41	33.97±10.78
Partially	29	32.37±7.43
No	60	40.90±13.81
Statistical Analysis		F=6.917. **p=0.001**
Informed about the Test Results of the Test		
Yes	94	35.76±12.21
Partially	24	36.70±12.37
No	12	45.25±9.5
Statistical Analysis		F=3.303. **p=0.04**

The Need for Counseling

More than half of pregnant women (61.5%) reported a need for genetic counseling after they received the test result (Table 4). When they were asked what kind of a counseling they would need, most of them claimed a comprehensive, detailed informative and psychologically supportive counseling was needed. The majority of those who had a positive test result (88.9%) expressed their need for genetic counseling ($p<0,05$). However, there was no significant association between the state anxiety scores of women who reported a need for counseling and those who didn't need counseling ($p>0,05$) (Figure 5).

Discussion

Increasing use of prenatal diagnostic methods and the perception and acceptance of these methods by couples brought its potential effects to the agenda. There are many factors affecting the agreement of pregnant women to participate to the genetic screening. In this study, all of the pregnant women accepted to participate in the genetic screening because of the approval of their physicians. Farrant [1985] points out that most of the women agreed to participate in the genetic screening because their obstetricians screen the anomalies; and those women consider that as a guarantee of the health of their fetus, in turn relieved themselves. Another factor effecting the agreement of women to participate in the screening program is the consideration of screening tests as a routine procedure [Green and Statham, 1999]. In this study, since only the women who had previously made their decision for amniocentesis had come to the genetic department for an interview, it was therefore not possible to examine the views of women who declined.

The first step in amniocentesis for women who agreed on prenatal screening is informed consent. As it is usual for all prenatal screening, the participation to amniocentesis must also be voluntary. Therefore, although prenatal screening tests are offered as standard and routine procedures, individuals have the right to decline. In Turkey, written informed consent for genetic testing and amniocentesis is a routine procedure.

However, in this study it was interesting to find out that 8.1% of pregnant women reported they were not asked for informed consent before amniocentesis. This finding raises the question on whether the consent form for amniocentesis was signed by women without comprehending its meaning, probably due to their concerns. O'Neill [2001] pointed out that the information provided by genetic tests may be so complex that it could be difficult to grasp intellectually, and accept emotionally, and in this sense, the issue of informed consent is a prepositional attitude in nature. O'Neill states that the consent given by someone for genetic testing who can not understand the implications of receiving a "positive result" cannot legitimate the action, in other words, consent obtained beyond the cognitive capacity of the individuals cannot be considered a true consent. In this respect, it is very important to get a written informed consent after informing the couples about amniocentesis in such a way that they can understand it clearly, by taking into consideration their education level [Pilnick and Dingwall, 2001].

Information relevant to amniocentesis can reduce the distress of pregnant women during the process of prenatal screening. As a matter of fact, the majority of the pregnant women in this study, reported psychological distress before genetic testing, even though they stated the concerns about the risks of amniocentesis (e.g. the potential of fetal injury, miscarriage) as the reason for their distress. Beeson and Golbus [1979], Black [1989] and Lavery et al. [2002]; reported in their studies that pregnant women who had undergone amniocentesis in the prenatal period experienced a high level of anxiety due to the concerns of procedure related complications and fear of fetal injury. Lavery et al. [2002] reported in their study that approximately half of the patients found the treatment process extremely stressful, some of them even expressed it as the most stressful event in their life. Marteau et al. [1988] pointed out in their study that the coping mechanisms of individuals with genetic amniocentesis depend on their knowledge on medical procedures, their anxiety levels before, during and after the procedures, and the communication between health professionals and patients. Other studies reported the stress perceived by individuals during the process of genetic testing and emphasize the need for genetic counseling [Robinson et al. 1989; Marteau and Croyle, 1998; Marteau et al., 1997; Michie et al., 1997; McConkie-Rosell et al., 1999; Pilnick and Dingwall, 2001; Esplen et al., 2001; Feeny et al., 2002, Hunter et al.,2005].

The trait anxiety experienced by pregnant women also contributes to their psychological distress. In this study, the trait anxiety level of pregnant women was found to be significantly higher than their state anxiety level. Many factors might influence the trait anxiety level of pregnant women, including their ethnic origin, education level, their perception of the economic status or income level, the status of employment, relationship between partners in marriage, the gender of the baby, the potential of fetal risk, having a previous child with anomaly, etc. Therefore, assessing the trait anxiety level of women is important in planning the education and managing health services for them.

There are also many factors that affect the state anxiety level of pregnant women during the process of prenatal screening, especially in an amniocentesis procedure. The major factors that increase the anxiety of women with regard to genetic amniocentesis are the indications for amniocentesis (e.g. advanced maternal age, positive history of spontaneously abortion, a previous child with abnormalities, abnormalities in triple screen tests). In particular, the concern of pregnant women about the risk of their fetus is the main contributing factor to their state anxiety. In the study of Yu et al. [1999] conducted on 23 mothers whose infants were at risk, the anxiety level of mothers with risky infants was found to be significantly higher than the mothers whose risk of infant was low.

In prenatal screening, the most important factor that increases the state anxiety level of pregnant women is a positive test result. In this study, 9 women were identified who had a positive genetic test result. The state anxiety level of women who had a positive test result was significantly higher than those who had a negative test result. Literature is replete of studies reporting that a positive test result can raise anxiety in women [Philips and Zinn, 1986; Van Zuuren, 1993; Ritvo et al., 2000; Pilnick and Dingwall, 2001; Esplen et al., 2001]. Evers-Kiebooms [1988], Green and Statham [1999] suggested that pregnant women are very concerned about the test results. Kallinen et al. [2001] identified a 5.9% abnormal test result of their cases, and they reported this as the reason for the anxiety experienced by the women. Esplen et al [2001] reported waiting for a test result and receiving a positive test result as the major cause of anxiety in couples. They suggested that knowing they were a carrier was also the main factor that elevates the anxiety of couples. This is especially important for couples who have a consanguinity marriage and who had a previous child with abnormalities. In this study, it was identified that 20.8% of participants had a consanguinity marriage. In consanguinity marriages, the similarity of genetic basis of couples may increase the risk of inheritance of autosomal recessive diseases for the child. In this case, the probability of having a child with autosomal recessive inherited disease, such as cystic fibrosis, FMF, Tay Sachs disease, Thalassaemia increases. In this study, the trait and state anxiety of women who had consanguinity marriages were found to be higher than those who did not. This finding can be helpful in understanding the association between some hereditary diseases and a consanguinity marriage.

A striking finding of this study was that the state anxiety level of pregnant women who had both a consanguinity marriage and a previous child with anomaly higher than those who didn't have a previous child with anomaly and consanguinity marriage. In addition, the proportion of having had a previous child was higher among those pregnant women who had a consanguinity marriage. Likewise, with regard to the association between the test results and having a previous child with anomaly, it was found that women who had a previous child with anomaly also received positive test results. All of these factors might be effective in increasing the anxiety level of pregnant women [Evers-Kiebooms, 1988; Denayer et al., 1992; Collins, 1999; Lavery et al., 2002]. In the study of Giarelli, [2003] it was found that individuals who were at genetic risk experienced more intense anxiety, depression, concern and the feeling of guilt. Weinans et al. [2000] found that women who had an abnormality in a child experienced a higher level of anxiety during the process of genetic amniocentesis. Ng et al. [2004] suggested that, even the consideration of women with a high risk of anomaly in their child increases their anxiety level. These results reveal the importance of considering pregnant women as special groups and providing them effective counseling on the process of pre-implantation, genetic screening, and on ovum donor or sperm using [Collins, 1999; Lavery et al., 2002]. However, donation of ovum or sperm is not legal in Turkey. In this respect, couples who had a family history of anomaly birth or whose likelihood of carrying is high should be referred to genetic counseling before marriage.

Another finding of this study was that almost all of those who received a positive test result wanted to terminate their pregnancy. An advantage of genetic amniocentesis is that it affords couples the opportunity to terminate an affected pregnancy [Kolker, 1989; Chescheir and Hansen, 1999; Tercyack et al., 2001]. However, it is worthwhile to emphasize that women confronted with an abnormal result in amniocentesis and genetic test result may feel confusion, anger and grief. Farrant [1985] suggests that pregnant women never prepare themselves for a positive test result. Denayer et al. [1992], and Evers-Kiebooms et al. [1993] in their studies with women who had undergone genetic amniocentesis, found that only one-fourth of women consider the termination of an affected pregnancy. The alternative to terminating an affected pregnancy may cause conflicting feelings in women [Wertz et al., 1994; Williams and Lessick, 1996; Lessick and Faux, 1998; McConkie-Rosell, et al., 1999; Burke et al., 2001]. Grant [2000] suggests that the decisions of couples on abortion is influenced by their beliefs and values, as well as their friends, families, media and the information provided by the health care personnel. On one hand, couples don't want to have an anomaly child; on the other hand, they may experience feelings of guilty, shame or loss of self-esteem. Consequently, a positive test result and a decision of abortion may have a physical, psychological, ethical or social impact, both on the fetus and the family.

One of the indications of amniocentesis is the previous experience of three or more spontaneous abortions. In the present study, the level of state anxiety in childless women with a previous experience of spontaneous abortion was found to be elevated. It is suggested that approximately 15% of all normal pregnancies results with spontaneous abortion and the incidence of chromosomal aberration in the abortion material is about 50% [Dagmar and Kalousek, 1999]. In this respect, it is very important to screen the genetic disorders caused by chromosomal aberrations, which can be identified by CVS or amniocentesis [Tabor et al., 1986; Schemmer and Johnson, 1993; Chescher and Hansen, 1999; Tercyack et al., 2001; Lavery et al., 2002; Fenny et al., 2002]. Previous experiences of a miscarriage increase the level of anxiety due to many factors. A bad obstetric history of women may raise anxiety and fear of pregnancy loss. The women may also be concerned about the potential of fetal injury during amniocentesis. Literature suggests that, although it is possible to detect fetal anomalies with amniocentesis, the complications of this procedure, which causes psychological trauma in women, include fetal loss in 0.5%, the risk of intra-amniotic infection in 0.1%, and amniotic fluid leakage in 1-2% [Tabor et al., 1986; Haris et. al., 2004]. Even though the level of risk seems to be low, pregnant women do not want to venture the risk and naturally experience distress. Similarly, the possibility to terminate the pregnancy if the test result is positive increases the anxiety of women.

The main indication for amniocentesis is advanced maternal age, which also increases the risk for chromosomal disorders. In this study the main reason for referring pregnant women to amniocentesis was also advanced age. Szczygiet and Kurpisz [2001], in their analysis of chromosomes XY, 13, 18 and 21 with women in advanced age, demonstrate that only 37.2% of embryos were normal. Therefore, it can be argued that women in age over 35 years are at risk for fetal anomaly, and literature also points out those women in advanced age are more anxious [Kolker and Burke, 1994; Haris et al., 2004; Hunter et al., 2005]. Since advanced age women are at the end of their reproductive periods and their ability for reproduction is reduced, this pregnancy might be considered by them as their last chance of having a baby. It is also suggested that women in advanced age experiences more anxiety because, they might have become pregnant following a long period of infertility treatments, or they might have had previous experiences of miscarriages or delayed their pregnancies due to various factors. However, Abuelo et al. [1991] found in their study that those patients with abnormal screening tests became more alarmed about the possibility of carrying an abnormal fetus than the women over 35, even though their statistical risks were similar. Several studies reported that women over 35 did not agree to participation in invasive procedures, such as genetic amniocentesis due to the risks involved, even though they have a positive test result [Fitzgerald et al., 1992; Browner et al., 1999].

Contrary to literature, the state anxiety level of young pregnant women (<26) was found to be higher than those in advanced age, which indicates that anxiety of women in advanced age was lower. Abuelo et al. [1991] found similar results to this finding. In this study women in the advanced age reported that they were not very anxious because genetic amniocentesis or any other screening tests are part of the routine diagnostic procedures. Therefore, women in young age should be assessed as risk groups and thus, counseling services toward these groups should consider their high anxiety level. The reasons for high anxiety among young women in the present study might be their perception of themselves as young and healthy women whose risk of having an anomaly baby should be lower than advanced age women, and their concerns on why such an invasive procedure is required for them. At the same time, most of those women in the younger age group might be more anxious since most of them were experiencing their first pregnancy, and had not such a previous experience. On the other hand, fear from the unknown or concerns because of invasive procedure risks might also have raised their anxiety.

Since genetic amniocentesis is an invasive procedure and pregnant women experience anxiety due to many factors, information related to amniocentesis genetic testing is very important. The protocols of screening are very complex and are increasingly changing and developing. Details, such as the number of markers to be analyzed and cut-off point for test positive may be different from one laboratory to another, even within the same laboratory at different times. Literature points out the importance of providing not only information to the couples that participate in the prenatal screening tests, but also the reports that information can reduce their anxiety [Abuelo et al., 1991; Ormond et al., 1996; Berkenstadt et al., 1999; Ekelin and Crang-Svalenius, 2004; Ng et al., 2004]. In this study, the state anxiety level of women who reported to be informed about the aim, procedure and potential of risks was found higher than those who reported they were not informed, and the difference between the two groups was statistically significant. In contrast to this finding, there are studies which reports that detailed information can actually raise anxiety [Lyoyd et al., 1996; Green and Statham, 1999].

Information that is understood and comprehended may be influenced by the education level [Pilnick and Dingwall, 2001]. This is important for nurses who are responsible for the education of individuals during the process of genetic testing. They should provide education and counseling according to the education level of individuals and answer the relevant questions. It is important to provide written information to individuals in genetic testing for conditions such as spina bifida, anencephaly, abdominal wall defects and chromosomal abnormalities. Participants of this present study reported their desire for more detailed information about the aim, procedure and risks as a result of genetic testing. This indicates the need of pregnant women in prenatal screening for comprehensive information and counseling before and after genetic testing and amniocentesis.

The state anxiety level of pregnant women according to their need for counseling after test results was high, but it was not statistically significant. Studies demonstrate that an effective counseling reduces the high level of anxiety [Marteau et al., 1991; Abuelo et al., 1991; Ormond et al., 1996; Michie et al., 1997; Berkenstadt et al., 1999; Pilnick and Dingwall, 2001; Ng et al., 2004]. However, in the study of Terycak et al. [2001] it was found that, prior to counseling, women's perceived risk was higher than actual risk. The education and counseling services for individuals should be planned and performed by taking into consideration their ability of understanding and comprehension.

Ormond and colleagues [1996] pointed out the importance of developing educational and informative material for genetic screening. Ng et al. [2004] reported that providing sufficient information and effective counseling before and after amniocentesis helps individuals in making appropriate decisions for them and reduces their level of anxiety. Hunter et al. [2005] also found that a session of 15-20 minutes in genetic counseling discussing the alternatives of prenatal tests improves the knowledge of individuals and helps them in decision making, but would not be effective in reducing their anxiety. Weinans et al. [2000] reported that the quality of counseling did not influence the decision of patients about amniocentesis, and the concern about the potential of information disclosure raised the anxiety of pregnant women. They also pointed out that prenatal counseling could only relieve mothers in the third trimester.

Consequently, the need for health professionals who can provide prenatal education and counseling for individuals has increased. Ideally, prenatal screening requires multidisciplinary teamwork. Nurses are the members of this health care team, who most frequently confront patients and their families, both in community health settings and various hospitals. Nurses play a vital role by identifying risk factors, evaluating individuals and their families, assessing family history in relation to any hereditary disease, providing genetic information and counseling, and referring individuals to genetic experts and services [Wertz et al., 1994; Haris and Haris, 1995; Jenkins, 2000; Middelton et al., 2002; Terzioglu and Dinç, 2004].

In conclusion, with the rapid developments in the field of genetics, the number of screening tests offered during the first and second trimester of pregnancy is increasing. These developments afford information about the health of the fetus. Meanwhile, social standards have changed, including greater expectations for a healthy child. Therefore, prenatal screening procedures are acknowledged as part of the routine procedures for the care of pregnant women. Amniocentesis, which affords detailed information about the health of a fetus, is one of the prenatal screening methods widely used. However, it has several serious complications that raise concerns of couples about the health of their baby. In the meantime, pregnant women are faced with difficult decisions depending upon the result of genetic tests. Feedback about the health of the fetus and the possibility for a positive test result increase the anxiety of women experienced during the process of prenatal screening. This in turn, indicates the physical, psychological and social effects of genetic amniocentesis. In this respect, it is important for health care members to be aware of the distress and concerns experienced by pregnant women, and to plan services to support them.

Limitations of this Study

This study has the following limitations.

1. The sample includes only pregnant women who came to the genetic department to receive their genetic test results after they had undergone amniocentesis and those who volunteered to participate in this study. As such, these results did not reflect the anxiety levels of those who prefer to be informed about their test results by telephone or fax or who did not come.
2. The reasons for genetic testing are heterogenic. Therefore, results of this present study are limited in generalizing to the subgroups.
3. No control group was recruited and patients after testing are not followed over time.
4. Lastly, the anxiety level of pregnant women might have been affected by the amniocentesis procedure and its risks alongside the genetic testing. Therefore, further studies that include large samples and focus on the anxiety assessment of pregnant women, both before the amniocentesis procedure and after genetic testing over time will be useful.

Acknowledgement

I would like to thank Leyla Dinç [Associated Professor in Hacettepe University School of Nursing] for her assistance in literature review and support.

References

[1] Abuelo, DN; Hopmann, MR; Barsel-Bowers, G; Goldstein, A. Anxiety in women with low maternal serum alpha-fetoprotein screening results. *Prenatal Diagnosis,* 1991, 11, 381-5.

[2] Astbury, J; Walters, WAW. Amniocentesis in the early second trimester of pregnancy and maternal anxiety. *Aust Fam Physician,* 1979, 8, 595-99.

[3] Beeson, D; Golbus, MS. Anxiety engendered by amniocentesis. *Birth Defects Orig Artic Ser,* 1979, 15, 191-7.

[4] Berkenstadt, M; Shiloh, S; Barkai, G; Katznelson, MB; Goldman, B. Perceived personal control (PPC): a new concept in measuring outcome of genetic counseling. *Am J Med Genet,* 1999, 82, 53-9.

[5] Black, RB. 1 and 6 month follow up of prenatal diagnosis patients who lost pregnancies. *Prenat Diagn,* 1989, 9, 795-804.

[6] Browner, CH; Preloran, HM; Cox, SJ. Ethnicity, bioethics, and prenatal diagnosis: the amniocentesis decisions of Mexican-origin women and their partners. *American Journal of Public Health,* 1999, 89, 1658-66.

[7] Burke, W; Pinsky, LE; Press, NA. Categorizing genetic tests to identify their ethical, legal, and social implications. *American Journal of Medical Genetics, (Semin.Med.Genet)* 2001, 106, 233-240.

[8] Cederlholm, M; Axelsson, O; Sjoden, PO. Women's knowledge, concerns and psychological reaction before undergoing an invasive procedure for prenatal karyotyping. *Ultrasound Obstet Gynecol,* 1999, 14, 267-72.

[9] Chescheir, NC; Hansen, WF. An update of current clinical practices in perinatology. *Pediatr Rev,* 1999, 20, 57-63.

[10] Collins, FS. Medical and Societal consequences of the Human Genome Project. *The New England Journal of Medicine,* 1999, 341, 28-37.

[11] Connor, JM; Ferfuson, MA. *Essential Medical Genetics.* Oxford: Blackwell Scientific Publications; 1993.

[12] Dagmar K; Kalousek MD. Clinical significance of a morphologic and genetic examination of spontaneously aborted embryos. *Frontiers in Fetal Health,* 1999, 1.

[13] Denayer, L; Evers-Kiebooms, G; De Boek, K; Van den Berghe, H. Reproductive decision making of aunts and uncles of child with cystic fibrosis: genetic risk perception and attitudes towards carrier identification and prenatal diagnosis. *Am J Med Genet,* 1992, 44, 104-11.

[14] Draper, H; Chadwick, R. Beware! Preimplantation genetic diagnosis may solve some old problems but it also raises new ones. *Journal of Medical Ethics,* 1999, 25, 114-120.

[15] Ekelin, M; Crang-Svalenius, E. Midwives' attitudes to and knowledge about a newly introduced foetal screening method. *Scand J Caring Sci,* 2004, 18, 287-93.

[16] Esplen, MJ; Madlensky, L; Mckinnon, W; Bapat, B; Wong, J; Aronson, M; Gallinger, S. Motivations and Psychosocial Impact of Genetic Testing for HNPCC. *American Journal of Medical Genetics,* 2001, 103, 9-15.
[17] Evers-Kiebooms, G; Swerts, A; Van den Bergha, H. Psychological aspects of amniocentesis: anxiety feelings in three different risk groups. *Clin. Genet,* 1988, 59, 196-200.
[18] Evers-Kiebooms, G; Denayer, L; Decruyenaere, M; Van den Berghe, H. Community attitudes towards prenatal testing for congenital handicap. *J Repro Infant Psychol,* 1993, 11, 21-30.
[19] Evers-Kiebooms, G; Swerts, A; Van den Berghe, H. Psychological aspects of amniocentesis: anxiety feelings in three different risk groups. *Clin Genet,* 1998, 35, 763-779.
[20] Farrant, W. *Who's for Amniocentesis? The Politics of Prenatal Screening*. In H Homans (ed.) The Sexual Politics of Reproduction. London, Gower, 1985, 96-177.
[21] Feeny, D; Townsend, M; Furlong, W; Tomkins, DJ; Robinson, GE; Torrance, GW; Mohide, PT; Wang, Q. Health-related quality-of life assessment of prenatal diagnosis: chorionic villi sampling and amniocentesis. *Genetic Testing,* 2002, 6, 39-46.
[22] Ferber, A; Onyeıje, CI; Zelop, CM; O'Reilly-Green, C; Divon, MY. Maternal pain and anxiety in genetic amniocentesis: expectation versus reality. *Ultrasound in Obstetrics and Gynecology,* 2002, 19, 35-44.
[23] Finley, SC. Genetic counselling. *Ala J. Med. Sci,* 1982, 19, 165-8.
[24] Fischer, RL; Biancilli, KW; Sehdev, H; Hediger, ML. Does light pressure effleurage reduce pain and anxiety associated with genetic amniocentesis? a randomized clinical trial. *The Journal of Maternal Fetal Medicine,* 2000, 9, 294-97.
[25] Fitzgerald, J; Streets, K; Prietst, J. Maternal serum screening risk for Down syndrome is considered positive if a triple-maker-derived risk exceeds the age-specific risk: Use of this definition for women aged 35 and over. *Prenatal Diagnosis*, 1992, 13, 152-53.
[26] Gardner, RJM; Sutherland, GR. Chromosome Abnormalities and Genetic Counselling : Oxford Monographs in Medical Genetics. Oxford: Oxford University Press; 1989.
[27] Giarelli, E. Safeguarding being: bioethical principle for genetic nursing care. *Nursing Ethics,* 2003, 10, 255-68.
[28] Gitzelman, R. Human genetic. *B. Medical Aspects*, 1982, 84, 425-36.
[29] Grant, SS. Prenatal genetic screening. *Online Journal of Issues in Nursing*, 2000, 5, Manuscript 3. Available http://www.nursingworld.org/ojin/topic13/tpc13_3.htm
[30] Green, JM; Statham, H. *Psychosocial Aspect of Prenatal Screening and Diagnosis.* İn: T Marteau, M Richards editors. The troubled helix: Social and Psychological Implications of the New Human Genetics. New York: Cambridge; 1999; 140-163.
[31] Haddow, JE; Palomaki, GE; Knight, GJ; Williams, J; Miller, WA; Johnson, A. Screening of maternal serum for fetal Down's Syndrome in the first trimester. *New England Journal of Medicine,* 1998, 338, 955-61.
[32] Harris, R; Harris, HJ. Primary care for patients at genetic risk. *British Medical Journal,* 1995, 311, 579-80.
[33] Harper, JC; Delhanty, JDA; Handyside, AH. *Preimplantation Genetic Diagnosis.* England: John Willy and Sons Ltd; 2001.

[34] Hancıoğlu, A; Tunçbilek, E; Onan, I. Türkiye'de Akraba Evlilikleri:[Consanguinity Marriage in Turkey]. 1998 Türkiye Nüfus ve Sağlık Araştırması. Papers in Demography. *Hacettepe University Institute of Population Studies*. Ankara, 2001.

[35] Haris, A; Monga, M; Wicklund, CA; Robbins-Furman, PJ; Strecker, MN; Doyle, NM; Mastrobattista, J. Clinical correlates of pain with amniocentesis. *American Journal of Obstetrics and Gynecology,* 2004, 191, 542-5.

[36] Hunfeld, JAM; Agterberg, G; Wyarilinoff, JW; Pasachier, J. Quality of life and anxiety in pregnancies after late pregnancy lose: a case control study. *Prenatal Diagnosis,* 1996, 16, 783-790.

[37] Hunfeld, JAM; Wladimiroff, J; Passcçhier, J; Veneme-Can Uden, MU; Frets, PG. Emotional reactions in women in late pregnancy (24 weeks or longer) following the ultrasound or severe or lethal fetal malformation. *Prenatal Diagnosis*, 1993, 13, 603-612.

[38] Hunter, AGW; Cappelli, M; Humphreys, L; Allanson, JE; Chiu, TT; Peeters, C; Moher, D; Zimak, A. A randomized trial comparing alternative approaches to prenatal diagnosis counseling in advanced maternal age patients. *Clin Genet,* 2005, 67, 303-13.

[39] Jeffords, JM; Daschle, T. Political issues in the Genome Era. *Science,* 2001, 291, 1249-1251.

[40] Jenkins, J. An Historical perspective on genetic care. *Online Journal of Issues in Nursing* 2000, 5, Manuscript2. Available: at: http://www.nursingworld/ojin/topic13/tpc13_2.htm.

[41] Jorgensen, FS. Attitudes to prenatal screening diagnosis and research among pregnant women with raised serum alpha fetoprotein: short term psychological effects. *J Psyshosom Res,* 1987, 31, 35-9.

[42] Juengst, ET. Human Genome research and the public interest: progress notes from an American Science Policy Experiment. *American Journal of Human Genetics,* 1994, 54, 121-128.

[43] Kallinen, J; Marin, K; Heinonen, S; Mannermaa, A; Palotie, A; Ryynanen, M. Wide scope prenatal diagnosis at Kuopio University Hospital 1997-1998: integration of gene tests and fetal karyotyping. *British Journal of Obstetrics and Gyneacology,* 2001, 108, 505-9.

[44] King, DS. Preimplantation genetic diagnosis and the 'new' eugenics. *Journal of Medical Ethics,* 1999, 25, 176-182.

[45] Kolker, A; Burke, BM. *Prenatal Testing: A Sociological Perspective.* Westport, Conn: Bergin and Garvey; 1994.

[46] Kolker, A. Advances in prenatal diagnosis: social-psychological and policy issues. *Int J Technol Assess Health Care,* 1989, 5, 601-17.

[47] Kowalcek, I; Muhlhoff, A; Bachmann, S; Gembruch, U. Depressive reactions and stress related to prenatal medicine procedures. *Ultrasound Obstet Gynecol,* 2002, 19, 18-23.

[48] Lavery, SA; Aurell, R; Turner, C; Castellu, C; Veiga, A; Bari, PN; Winston, RM. Preimplantation genetic diagnosis: patients' experiences and attitudes. *Human Reproduction*, 2002, 17, 2464-70.

[49] Lessick, M; Faux, S. Implications of genetic testing of children and adolescents. *Holistic Nursing Practice,* 1998, 12, 38-46.

[50] Lyoyd, S; Watson, M; Waites, B; Meyer, L; Eeles, L; Ebb, SS; Tylee, A. Familial breast cancer: a controlled study of risk perception, psychological morbidity and health beliefs in women attending for genetic counseling. *Br J Cancer,* 1996, 74, 482-7.
[51] Marteau, TM; Johnston, M; Plenicar, M; Shaw, RW; Slack, J. Development of a self administered questionnaire to measure women's knowledge of prenatal screening and diagnostic testing. *J Psychosom Res,* 1988, 32, 403-8.
[52] Marteau, TM; Croyle, R. Psychological responses to genetic testing. *British Medical Journal,* 1998, 316, 693-696.
[53] Marteau, TM; Dundas, R; Axworthy, D. Long term cognitive and emotional impact of genetic testing for carriers of cystic fibrosis: the effects of gender and test result. *Health Psychol,* 1997, 16, 51-62.
[54] McConkie-Rosell, A; Spiridigliozzi, GA; Rounds, K; Dawson, DV; Sullivan, JA; Burgess, D; Lachiewicz, AM. Parental Attitudes Regarding Carrier Testing in Children at Risk for Fragile X Syndrome. *American Journal of Medical Genetics,* 1999, 82, 206-211.
[55] Michie, S; Marteau, TM; Bobrow, M. Genetic counselling: the psychological impact of meeting patients' expectations. *J Med Genet,* 1997, 34, 237-241.
[56] Middelton, L; Dimond, E; Calzone, K; Davis, J; Jenkins, J. The role of the nurse in cancer genetics. *Cancer Nursing,* 2002, 25, 196-206
[57] Ng, CCM; Lai, FM; Yeo, GSH. Assessment of maternal anxiety levels before and after amniocentesis. *Singapure Med J,* 2004, 45, 370-74.
[58] O'Brien, WF. Midtrimester genetic amniocentesis. a review of fetal risks. *J Reprod Med,* 1984, 29, 59-63.
[59] O'Neill, O. Informed Consent and Genetic Information. *Stud. Hist. Phil. andBiomed Sci,* 2001, 32, 689-704. Available at: http://www.elsevier.com/locate/shpsc Accessed on 30 January 2003
[60] Öner, N; Le Compte, A. Durumluk ve Sürekli Kaygı Envanteri El Kitabı [Handbook of State and Trait Inventory]. İstanbul: Boğaziçi Üniversitesi Yayınları; 1982.
[61] Ormond, KE; Pergament, E; Fine, BA. Prescreening education in multiple marker screening programs: the effect on patient anxiety and knowledge. *J Genet Couns,* 1996, 5, 69-80.
[62] Paabo, S. The human genome and our view of ourselves. *Science Magazine,* 2002, 16, 1219-1220.
[63] Phipps, S; Zinn, AB. Psychological response to amniocentesis. effects of coping style. *Am J Med Genet,* 1986, 25, 143-8.
[64] Pillitteri, A. *Maternal And Child Health Nursing.* Philadelphia; JB. Lippincott Company; 1992: 157-158.
[65] Pilnick, A; Dingwall, R. Research directions in genetic counselling: a review of the literature. *Patient Education and Counseling,* 2001, 44, 95-105.
[66] Qulligan, EJ; Zuspan, FP. *Current Therapy in Obstetrics and Gynecology.* 5th edition, Saunders Company, 2000.
[67] Raines, DA. Ethical implications of genetic testing. *Nursing Clinics of North America,* 1998, 33, 275-286.
[68] Ritvo, P; Robinson, G; Irvine, J; Brown, L; Matthew, A.; Murphy, KJ; Stewart, DS; Styra, R; Wang, C; Mullen, M; Cole, D; Rosen, B. Psychological adjustment to familial

genetic risk assessment: differences in two longitudinal samples. *Patient Education and Counseling,* 2000, 40, 163-172.
[69] Robinson, GE; Garner, DM; Olmsted, MP; Shime, J; Hutton, EM; Crawford, BM. Anxiety reduction after chorionic villus sampling and genetic amniocentesis. *Am J Obstet Gynecol* , 1989, 1859, 953-6.
[70] Robinson, GE; Johnson, JAM; Wilson, RD; Gajjar, M. Anxiety reduction after early and mid tirmester prenatal diagnostic testing. *Arch Womens Ment Health,* 1998, 1, 39-44.
[71] Roth, MT; Painter, RB. Genetic discrimination in health insurance. *Nursing Clinics of North America,* 2000, 35, 731-755.
[72] Saraçoğlu, F. *Fetal Tanı ve Tedavi.* Ankara: Güneş Kitabevi; 1998.
[73] Schemmer, G; Jhonson, A. Genetic amniocentesis and chorionic villus sampling. *Obstetrics and Gynecology Clinics Of North America,* 1993, 20, 497-521.
[74] Sjogren, B; Uddenberg, N. Prenatal diagnosis and psychological distress: amniocentesis or chorionic villus biopsy. *Prenatal Diagnosis,* 1989, 9, 477-487.
[75] Sjogren, B. Psychological indications for prenatal diagnosis. *Prenatal Diagnosis,* 1996, 16, 449-54.
[76] Szczygiet, M; Kurpisz, M. Chromosomal anomalies in human gametes and pre implantation embryos and their potential effect on reproduction. *Andrologia,* 2001, 33, 249-65.
[77] Tabor, A; Jonsson, MH. Psychological impact of amniocentesis on low risk women. *Prenatal Diagnosis*, 1987, 7, 443-9.
[78] Tabor, A; Philip, J; Mansen, M. Randomized controlled trial of genetic amniocentesis in 4606 low risk women. *Lancet,* 1986, 7 [8493], 1287-93.
[79] Tercyak, KP; Johnson, SB; Roberts, SF; Cruz, AC. Psychological response to prenatal genetic counseling and amniocentesis. *Patient Education and Counseling,* 2001, 43, 73-84.
[80] Terzioglu, F; Dinc, L. Nurses' views on their role in genetics. *JOGNN*, 2004, 33, 756-64.
[81] Tinkle, MB; Cheek, DJ. Human genomics: challenges and opportunities. *JOGNN,* 2002, 31, 178-187.
[82] Van Zuuren, FJ. Coping style and anxiety during prenatal diagnosis. *J Reprod Infant Psychol,* 1993, 11, 57-9.
[83] Vastag, B. Experts wrestle with social, ethical implications of human genome research. *JAMA,* 2001, 285, 721-722.
[84] Yu, MS; Norris, JM; Mitchell, CM; Butler-Simon, N; Groshek, M; Follansbee, D; Erlich, H; Rewers, M; Klingensmith, GJ. Impact on maternal parenting stress of receipt of genetic information regarding risk of diabetes in newborn infant. *Am J Med Genet,* 1999, 86, 219-26.
[85] Weinans, MJ; Huijssoon, AM; Tmystra, T; Gerrits, MC; Beekhuis, JR; Mantingh, A. How women deal with the results of serum screening for Down syndrome in the second trimester of pregnancy. *Prenatal Diagnosis,* 2000, 20, 705-8.
[86] Wertz, DC; Fanos, JH; Reilly, PR. Genetic testing for children and adolescents. who decides? *JAMA*, 1994, 272, 875-881.
[87] Williams, JK; Lessick, M. Genome research: implications for children. *Pediatric Nursing,* 1996, 22, 40-46.

[88] Turkey Demographic and Health Survey 2003; Hacettepe University Institute of Population Studies, Ankara.

In: Focus on Genetic Screening Research
Editor: Sandra R. Pupecki, pp. 57-71
ISBN 1-60021-229-8

Chapter III

Mendelian Disorders among Jews and Population Screening for Reproductive Purposes

Joël Zlotogora[*]
Department of Community Genetics, Public Health Services,
Ministry of Health, Israel

Abstract

The Jewish population has remained isolated for centuries, and as a direct result, many genetic disorders are relatively frequent among Jews. For most of the severe autosomal recessive diseases no treatment is available and prevention is therefore the only means to reduce the morbidity. The Tay - Sachs disease carrier screening prevention program was initiated in 1971 among Jews of Ashkenazi origin and remained up to now very successful. In the last decade, the possibility of carrier detection has also become available for most of the other relatively frequent disorders existing in the different Jewish communities, leading to the possibility of expanding genetic screening programs.

According to the guidelines for population screening first delineated by Wilson and Jungner for the World Health Organization, the three major criteria for genetic screening for reproductive purposes are (1) the disease is severe and occurs predominantly in a well-defined population, (2) the detection of heterozygotes is possible using simple, accurate and inexpensive technology and (3) it is possible to prevent the birth of affected children. While the application of these criteria seems easy, the definitions of severity and frequency may be controversial. For instance, the tests are often offered to couples of mixed origin even though the risk is very low. In addition, the genetic screening offered to the Jewish population includes disorders such as Gaucher disease or deafness, which for many are not considered severe enough to justify population screening.

[*] Address for correspondence: Joel Zlotogora MD, Ph.D. Department of community Genetics; Public Health services, Ministry of Health. Building 67, Sheba Medical center; TEL HASHOMER; RAMAT GAN. 52621 ISRAEL; Tel 972 2 5348432; FAX 972 3 5355166

A screening program aimed at the Jewish population should be based upon criteria accepted by the communities and must be preceded by genetic counseling, emphasizing its aims and limitations.

1. Introduction

The Jewish people have remained a distinct entity for centuries, mainly by keeping marriages within the community. After its dispersion, many different communities were created, often isolated from one another. As a direct result of these events many genetic disorders have been found with an increased frequency among Jews, with marked differences between the different communities.

Mendelian disorders, relatively frequent among Jews, may be classified according to the mode of inheritance and clinical presentation [1,2,3]. Most of these disorders are autosomal recessive and present with different degrees of severity and morbidity. Some diseases are lethal in early childhood and/or lead to severe disability other are mild almost without clinical symptoms and in a third group there is some degree of disability, often with different degrees of suffering from one patient to another. The last group includes disorders with late onset.

Since, for most of the severe autosomal recessive diseases no treatment is available, prevention is the only means to reduce the morbidity. The Tay-Sachs disease carrier screening prevention program was initiated in 1971 [4] among Jews of Ashkenazi origin and has been successful. With the advances in human genetics, the molecular basis of many relatively frequent disorders present among the different Jewish communities has been characterized allowing for a possible expansion of the screening program for reproductive purposes.

2. The Jewish People

The Jewish people originate from the Middle East. After the destruction of the First Temple, part of the community remained in Palestine while another part moved eastward founding the Babylonian Jewry, ancestors of the Iraqi and Iranian Jews. With the rise of the Greco-Roman empires, Oriental Jews also began to move westward as far as Spain and Portugal where a large Jewish community developed in the Middle Ages. Sephardic Jews (from the Hebrew word for an area often identified with Spain) are descendants of the Jews who were forcibly expelled during the Inquisition in Spain and settled mostly in the countries along the Mediterranean Sea and in the Netherlands and the New World. The major movement of Oriental Jews toward Eastern Europe was in the Middle Ages to France and Germany. Later on, there were two important waves of migrations of the Ashkenazi Jews (from a word meaning Germany) in central Europe: the first in the 15th-16th century eastward to Bohemia and eventually to Poland-Lithuania and then from the end of the 18th century back to the west (Germany, Netherlands, England, USA).

While differences existed between the various regions where the Ashkenazi Jews were living, there are no true subgroups among them. On the other hand, most of the Oriental and Sepharadic Jewish communities remained geographically separated and developed as distinct identities. The non-Ashkenazi communities represented more than 90% of the Jews in the

12th century and today some 75% of them are of Ashkenazi origin. In 2004 there were 5.5 million Jews living in Israel, approximately 50% were of Ashkenazi origin.

Consanguineous marriages are allowed in the Jewish religion and they were common in all the communities including the Ashkenazi Jews. In a survey in Israel after the foundation of the State, the consanguinity rate varied from 1.4% among Ashkenazi Jews to 28.7 % among Jews from Iraq [5]. In the last decades, inter-community marriages are more and more frequent and within each of the communities consanguineous marriages are declining rapidly [6].

3. Relatively Frequent Severe Autosomal Recessive Disorders among the Jews

Ashkenazi Jews

Today the largest Ashkenazi Jewish communities are in the United States and in Israel. Among the frequent severe diseases among Ashkenazi Jews are Tay-Sachs disease, Familial dysautonomia and Cystic fibrosis, with a carrier frequency between 1:25 and 1:30 and Canavan disease with a carrier frequency of 1:60. Among Ashkenazi Jewish patients affected with Tay-Sachs disease three major mutations in the alpha subunit of beta hexosaminidase were identified [7]. The two most frequent mutations are: [HEXA, 4-BP INS, EX11] (73% of the mutant alleles) and [HEXA, IVS12+1G>C], (17.6% of the mutant alleles). The third mutation [HEXA, G269S], which is relatively rare (3.5% of the mutant alleles) is responsible for the adult type of Tay-Sachs disease. The prevalence of Cystic fibrosis among Ashkenazi Jews, almost similar to the one in Caucasian populations, however, the distribution of the different mutations in the CFTR gene is unique. Among Ashkenazi Jews, 5 mutations represent 97% of the mutations found among the patients [CFTR, W1282X] (49%), [CFTR, ΔF508] (31%), [CFTR, G542X], [CFTR, N1303K] and [CFTR, 3849+10Kb C→T] [8]. Familial dysautonomia is a disease due to a disturbance of autonomic and peripheral sensory functions, which is the only other disorder present almost exclusively among Ashkenazi Jews. Two mutations have been found as a frequent founder mutation [IKBKAP IVS20 +6T>C)] and a rare mutation [IKBKAP R696P] [9]. Canavan disease is a rare degenerative disease of the brain due to the deficiency of the enzyme aspartoacylase in which one major mutation [ASPA, E285A] accounts for 83% of the alleles among Ashkenazi Jews in the USA and all the alleles in Israel [10]. Another mutation [ASPA, Y231X] is relatively frequent among Ashkenazi Jews living in the USA (15% of the alleles).

Other severe diseases are much less frequent, such as: Bloom syndrome, Fanconi anemia, Niemann Pick disease type A, Mucolipidosis IV, lipoamide dehydrogenase deficiency, Glycogenosis Ia, MSUD, Nemaline myopathy and Usher syndrome. In Bloom syndrome the most prominent clinical symptoms are the severe pre- and post-natal growth retardation and an unusually high rate of neoplasia. Most of the patients reported in the world who are affected with Bloom syndrome are Ashkenazi Jews and among them a unique mutation [BLM, 6-BP DEL/7-BP INS] is found [11]. In Fanconi anemia congenital malformations radial defects in particular, are common, however, the major problem is the marrow failure that affects all the elements resulting in anemia, thrombocytopenia and leucopenia. Among

Ashkenazi Jews, a single mutation [FANCC, IVS + 4 A to T] is found [12]. Niemann Pick disease type A occurs in infants and t is characterized by jaundice, enlarged liver, and profound brain damage. Affected children rarely live beyond 18 months. Three mutations with similar frequency account for more than 95% of the mutant alleles found among Ashkenazi Jewish patients: [SMPD1, R496L], [SMPD1, L302P] and [SMPD1, fsP330] [13]. Mucolipidosis IV is a disorder characterized by psychomotor retardation and ophthalmological abnormalities, including corneal opacities, retinal degeneration, strabismus and often myopia. Among the two founder mutations: one [MCLOM1, IVS3 1+A>G] represents 75% of the alleles, the other [MCLOM1, del exon 3-7] is rarer [14]. Lipoamide dehydrogenase deficiency is characterized by recurrent attacks of vomiting, abdominal pains and encephalopathy with abnormal liver function with variable age of onset [15]. Two mutations in the LAD gene have been found among Ashkenazi patients. The mutation [LAD, G229C] is the most frequent, while the [LAD, Y35X] is rare. Glycogenosis Ia usually manifests itself in the first year of life by severe hypoglycemia, hepatomegaly, growth deficiency and bleeding diathesis. One major mutation is prevalent among Ashkenazi Jews [G6PC, R83C] [16]. Maple Syrup Urine Disease (MSUD) is an inherited metabolic disorder that, if untreated, causes mental retardation, physical disabilities and death. The classical form of the disease has been diagnosed in several Ashkenazi Jewish patients in the newborn screening in New York due to one founder mutation [BCKDHB, R183P] [17]. Nemaline Myopathy is a rare inherited neuromuscular disease that is usually apparent at birth and characterized by extreme muscle weakness. The disease was reported in several Ashkenazi Jewish families because of a founder mutation [NEB, 2,502-BP DEL] [18]. Usher syndrome type 1 is characterized by profound congenital deafness, vision problems starting in early childhood, and severe balance problems. A single mutation in the gene [PCDH15, R245X] is frequent among Ashkenazi Jews with the syndrome [19].

Two other disorders Gaucher disease and non-syndromic deafness are not considered as severe and lead to variable disability are particularly important since they are frequent among the Ashkenazi Jews.

In the adult type of Gaucher disease, most of the patients are asymptomatic or affected with a very mild disease. In only a small proportion of the patients, the disease is severe and in those cases enzymatic treatment is available [20]. The mutation [GBA, N370S] is present in 85% of the patients and the other mutations [GBA, 84GG] (6%) and [GBA, L444P] (3.5%) and [GBA, IVS2, +1G>A] (1%) are rarer, but usually associated with a more severe clinical picture. In non-syndromic deafness, two frame-shift mutations in the GBJ2 gene encoding for connexin 26 are frequent among Ashkenazi Jews [21]. One of these mutations [GJB2 167delT] is present in approximately 4% of the Ashkenazi Jews. Another deletion mutation in GJB6 coding for connexin 30 was found to be also frequent leading to deafness in homozygosity or in cooperation with mutations in connexin 26 [22].

Jews from North Africa

In each of the countries of North Africa, Morocco, Algeria, Tunisia and Libya an ancient Jewish community existed for more than 2,000 years. Today the largest communities, including Jews originating from the different North African countries, are in Israel and France.

Among the severe genetic disorders reported among Jews from North Africa, ataxia telangiectasia, glycogen storage disease type III, cerebrotendinous xanthomatosis, Tay-Sachs disease, adrenal hyperplasia IV, Fanconi anemia and cystinosis are frequent, particularly among Moroccan Jews.

The predominant clinical features of ataxia telangiectasia are neurologic symptoms and telangiectasias that typically appear after the age of 2 years. Variable immunodeficiency is also found in these patients and may lead to recurrent upper respiratory and lung diseases. In addition, the patients with ataxia telangiectasia have an increased risk for different types of cancer, in particular, lymphomas as a result of chromosomal instability. A single mutation [ATM, R35X] was found in almost all the Jewish patients from North Africa [23]. The patients affected with Glycogen storage disease type III usually present in childhood a symptomatology similar to glycogen storage disease type I, however, in some cases, there is a spontaneous improvement at puberty. A founder mutation [AGL, 4,455delT] was found in all North African Jewish patients examined [24]. Cystinosis. The clinical picture is dominated by the symptomatology related to the renal Fanconi syndrome. The founder mutation [CTNS, G339R] is responsible for most of the cases [25]. In Fanconi anemia congenital malformations radial defects in particular, are common, however, the major problem is the marrow failure. Among North African Jewish patients, most of the alleles are [FANCA, 2172-2173 INS], the other are [FANCA, 4275delT] [12]. Adrenal hyperplasia IV is a disorder in which the clinical symptomatology is variable, related to the androgen and mineralocorticoid excess secondary to the deficiency of the steroid 11 ß-hydroxylase. Among the Moroccan Jewish patients a single mutation [CYP11B1, R448H] was identified in almost all the alleles and since most families originated from the Atlas Mountains in Morocco it is probable that the founder was a Berber Jew [26]. In cerebrotendinous xanthomatosis juvenile cataracts are often the presenting symptom while tendon xanthomas often develop later in the third and fourth decade. Most patients have low intelligence or are mentally retarded and develop progressive spasticity with ataxia. The disease is prevalent only among Moroccan Jews and among them, two mutations are frequent: [CYP27, IVS4+1G>A] and [CYP27, 1-BP DEL, FS] [27]. Tay-Sachs disease of the infantile type is less frequent among Moroccan Jews than among the Ashkenazi Jews and in both cases it is due to multiple mutations mostly different between these communities. Among the Moroccan Jewish patients 7 different mutations have been identified, three of which were frequent [28].

Among the Jews originating from Libya, vacuolating megalencephalic leukoencephalopathy is frequent because of a founder mutation [MLC1, G59E] [29]. The disease is characterized by accelerated head growth and neurologic involvement. Most develop a seizure disorder and a severe cerebral white matter change on neuroimaging. The progression is very slow and the cognitive functions remain almost intact. A limb girdle muscular dystrophy with muscular weakness that appears in the second half of the third decade as gait difficulties is also frequent among Jews from Libya due to a founder mutation [DYSF, 1-BP DEL] [30].

Jews from Iraq and Iran

The Jewish communities of Iraq and Iran are among the oldest and trace their history back to 6th century b.c.e., the time of the First Temple. The two communities have a common history in the pre-Islamic period and thereafter were mostly distinct. After the legalization of immigration from Iraq in 1950, most of the Iraqi Jews emigrated to Israel. Jewish immigration from Iran has been mainly since the creation of the state of Israel and on the eve of the Islamic Revolution to Israel as well as other destinations, in particular the USA, where a relatively large Iranian Jewish community now lives.

Some of the relatively frequent disorders are found both among Jews from Iran and Iraq, such as Congenital familial myasthenia gravis and thalassemia. Congenital familial myasthenia gravis is a disorder in which the most common manifestations are ptosis, limited facial expression with a myopathic appearance, together with mandibular prognathism, malocclusion and open bite. All the patients from Iran and Iraq are homozygous for the same founder mutation [RAPSN, -38A>G] [31].

Among the more frequent disorders found among Iraqi Jews, some are quite severe, such as 3-Methyl glutaconic aciduria and Glanzmann thrombasthenia.

3-Methyl glutaconic aciduria was first described among Iraqi Jews and includes bilateral optic atrophy and neurological signs. The visual impairment and decreased distance acuity does not seem to be progressive. Patients present with neurological signs related to the extrapyramidal system and most have a low borderline intelligence up to mild mental retardation. All patients excrete large amounts of 3-methyl glutaconic acid in the urine. The high frequency is due to a founder mutation [OPA3, G>C intron 1 acceptor] [32]. Glanzmann thrombasthenia is a disorder of platelet function manifested by severe bleeding tendency. The disease may be caused by mutations in either glycoprotein IIb or IIIa genes. Among the Iraqi Jews the incidence is estimated to be 1:7,700 and two mutations have been characterized [33]. The most frequent mutation is [ITGB3, 11-BP DEL, EX12] (85% of the alleles). Another mutation also found in this community is a 11.2Kb deletion between intron 9 and exon 14 leading to a shift in the reading frame and a stop codon (less than 8% of the alleles).

Among the disorders frequently found in Iranian Jews some are severe, such as Inclusion body myopathy, and Autoimmune Polyendocrinopathy-Candidiasis-Ectodermal Dystrophy (APCED).

Inclusion body myopathy is a disease in which muscular weakness usually appears in the third decade as gait difficulties. The progression is gradual but the quadriceps stay strong, even in the advanced stage of the disorder, allowing patients to stand and walk until late in the course of the disease. While most known patients are Iranian Jews, a few originate from Afghanistan and Egypt. The high frequency is due to a founder mutation [GNE, M712T] [34]. Autoimmune Polyendocrinopathy-Candidiasis-Ectodermal Dystrophy (APCED) is relatively frequent among the Iranian Jews due to a founder mutation [AIRE, Y85C] [35]. The clinical picture is variable and progressive, but the cardinal features are hypoparathyroidism, adrenal insufficiency and chronic mucocutaneous candidiasis. The gene was characterized, but the molecular basis of the disease among the Iranian Jewish patients has not yet been identified.

Jews from Kurdistan

An isolated ancient Jewish community has been living in the mountains of Kurdistan and included three separate subgroups: the largest in Northern Iraq, the second largest in Iran and the smallest in eastern Turkey. In 1950, almost all the communities immigrated to Israel.

Thalassemia is particularly frequent among Kurdish Jews with an estimated carrier frequency of 15-20% for ß-thalassemia. For ß-thalassemia, 13 mutations in the ß-globin gene were identified among the Kurdish Jews, 4 of which were characterized for the first time in this population: [HBB, TRP37FS], [HBB, C88A], [HBB, PRO36FS], [HBB, 3-UNT, A-G, +6] [36]. There was very little overlap in the mutations found in the three communities, further demonstrating the isolation of the three communities even one from the other, in particular because of the geographic conditions. It seems that ß-thalassemia was frequent among the Kurdish Jews, both because of multiple founder effects and the selective advantage of the heterozygotes against malaria, which was prevalent in the region.

Jews from Yemen

The Jews of Yemen trace their origins back to biblical times; the community probably arose from the settlement of Jewish traders and merchants. Several disorders are found with a relatively high frequency in the community, in each of them because of a founder effect.

Phenylketonuria is found with an incidence of 1:5000 among Jews from Yemen. All the families of the affected patients originated from San'a in which a common ancestor lived in the 18th century [37]. The possibility that the increased incidence of the disease in this isolated population is the result of a founder effect is supported by molecular studies, which demonstrated a similar deletion in the gene of all the patients. Metachromatic leukodystrophy is a rare neuro degenerative disease secondary to the deficiency of the lysosomal enzyme arylsulfatase A. The disease is also relatively frequent in Jews from Yemen and all the Jewish patients from the region are homozygous for a same mutation [ARSA, P377L] [38].

4. Population Screening for Reproductive Purposes among the Jews

Guidelines for population screening were first delineated by Wilson and Jungner for the World Health Organization and have been subsequently modified [39]. The three major criteria for genetic screening for reproductive purposes are (1) the disease is severe and occurs predominantly in a well-defined population, (2) the detection of heterozygotes is possible using a simple, accurate and inexpensive technology and (3) it is possible to prevent the birth of affected children. Looking at these criteria, the most problematic are the definitions of severity and frequency. Tay-Sachs disease that is lethal, frequent among the Ashkenazi Jews and for which a simple test for heterozygote detection was available was among the first diseases fitting the criteria [4]. While there is a consensus that among Jews several diseases like Tay-Sachs are both severe and frequent, several relatively frequent diseases are either milder or more rare (Table 1). With the characterization of the molecular basis of almost all the genetic diseases relatively frequent among the Jews, there are some who will argue that if a test is already performed to detect carriers of relatively frequent diseases such as Tay-Sachs, then it should also be used to diagnose carriers of rare disorders in which such testing is possible. However, it should be remembered that even when the rate of detection is optimal there are many causes for false positive/negative results and in particular for very rare diseases the overall possibility of an error will be higher than the one to find a carrier. Another problem that should be considered is that while many disorders were found to be relatively frequent in the different Jewish communities, in Israel there is an increasing tendency of marriages in which each spouse comes from a different Jewish community. It is expected that rapidly a large proportion of the Jewish population in Israel will be of mixed origin [6]. Accordingly, the carrier frequency for those diseases which were limited to one community begun to decrease while, on the other hand the size of the population at risk is increasing, leading to new carrier frequencies of genetic diseases. For instance, while the carrier frequency of TSD is 1/30 among Ashkenazi Jews representing approximately 50% of the Jewish population in Israel, mixed marriages will create a new carrier frequency among Israeli Jews around 1/60. In the same time for disorders, which are relatively frequent only in small communities, the carrier frequency in the general Jewish population will be much lower. For example, the carrier frequency of metachromatic leukodystrophy, which is 1/50 among Jews from Yemen, will become as low as <1/500 in a mixed Jewish population. Therefore, for those disorders which are frequent in relatively small communities or those present in larger communities in which the frequency is not so high, the mixing of the population will lead to frequencies which do not justify screening.

Since for the majority of genetic diseases there is no possible cure, the goal of a carrier screening is prevention. The most effective way to reduce the impact of these diseases is either by primary prevention, namely by using the screening for premarital counseling or by secondary prevention, e.g. giving the potential couples at risk the accessibility for prenatal diagnosis and/or interruption of the pregnancy if the fetus is affected. One example is the screening performed among most of the couples of Ultra Orthodox Jewish community, which is a successful program aimed at the primary prevention of genetic diseases by detecting the couples at risk before the engagement and therefore preventing marriages between heterozygotes [40]. This kind of genetic screening was adapted by the community members with cooperation and blessing of its religious leaders, since it has met both their needs and culture. The two main reasons for the program's acceptability are that the majority of the marriages among the Ultra Orthodox Jewish community are prearranged and the termination of a pregnancy is unacceptable.

Table 1. Autosomal recessive Mendelian disorders relatively frequent among Jews according to their origin. The more severe and relatively frequent diseases are in bold

Disorder	MIM	Heterozygote frequency
ASHKENAZI JEWS		
Abetalipoproteinemia.	200100	Rare
Adrenal hyperplasia III Non classical 21 OH deficiency	201910	1:6
Bloom syndrome	**210900**	**1:94-112**
Canavan disease	**271900**	**1:41-60**
Cystic fibrosis	**219700**	**1:25**
Deafness	121011	1:25
Dysautonomia, familial	**223900**	**1:30-36**
Factor XI (PTA) deficiency	264900	1:12
Familial Mediterranean Fever	249100	1:5 – 1:10
Fanconi anemia	**227645**	**1:77-218**
Gaucher disease type I	230800	1:7-18
Glycogenosis Ia	**232200**	**1:71**
Glycogenosis IV (adult)	232500	Rare
GlycogenosisVII Phosphokinase deficiency, late onset	232800	Rare
Hermansky-Pudlak syndrome	203300	1:235
Lipoamide dehydrogenase deficiency	**246900**	**1:94**
MSUD	**248600**	**1:113**
Mucolipidosis IV	**252650**	**1:100-112**
Nemaline myopathy		**1:108**
Niemann-Pick disease A	**257200**	**1:90**
Pentosuria	260800	1:35
Persistent hyperinsulinemic hypoglycemia of infancy	600509	1:125-160
Tay-Sachs disease	**272800**	**1:25-30**
Thalassemia, alpha	141800	1:17
Usher syndrome	**602083**	**1:72**
JEWS FROM MOROCCO		
Adrenal hyperplasia, 11 beta hydroxylase	**202010**	**1:30**
Albinism, oculocutaneous	203100	1:30
Ataxia telangiectasia	**208900**	**1:80**
Complement C7 deficiency	217070	1:100
Cystinosis	**219800**	**1:100**
Cerebrotendinous xanthomatosis	**213700**	**1:70**
Dubin Johnson syndrome	237500	1:100
Factor VII	227500	1:42
Familial Mediterranean Fever	249100	1:7
Fanconi disease	**227650**	**1:100**
Glycogen storage disease III	**232400**	**1:35**
Tay-Sachs disease	**272800**	**1:110**
JEWS FROM TUNISIA		
Brittle cornea syndrome	229200	unknown
Factor V and VIII deficiency	227300	unknown
Familial Mediterranean Fever	249100	1:7
Selective intestinal malabsorption B12	261100	Rare
Mucolipidosis III	252600	unknown

Table 1. Continued

Disorder	MIM	Heterozygote frequency
JEWS FROM LIBYA		
Cystinuria	600918	1:25
Limb girdle muscular dystrophy	**253601**	**1:10**
Familial Mediterranean Fever	249100	1:5
Megaencephalic leukoencephalopathy		**1:50**
JEWS FROM ALGERIA		
Familial Mediterranean fever	249100	1:13
JEWS FROM EGYPT (Karaites)		
Spinal Muscular Atrophy I	253300	1:10
JEWS FROM IRAQ		
Factor XI deficiency (PTA deficiency)	264900	1:30
Familial Mediterranean Fever	249100	1:15
Type III 3-Methyl glutaconic aciduria	**258501**	**1:10**
Microphthalmia/ anophthalmia	251600	unknown
Myasthenia gravis, infantile	254210	unknown
Pituitary dwarfism II (Laron)	262500	unknown
Thrombasthenia (Glanzmann)	**273800**	**1:40**
JEWS FROM IRAN		
Colour blindness, total (achromatopsia)	216900	unknown
Corticosterone methyl oxydase II deficiency	124080	1:30
Dubin Johnson syndrome	237500	1:20
Factor V and VIII deficiency	227300	unknown
Factor VII deficiency	227500	1:40
Microphthamia/anophthalmia	251600	unknown
Myasthenia gravis, infantile	254210	unknown
Myoneurogastrointestinal encephalopathy	603041	Unknown
Inclusion Body Myopathy	**600737**	**1:12**
Pituitary dwarfism II (Laron)	262500	unknown
Polyglandular deficiency syndrome	**240300**	**1:50**
JEWS FROM TURKEY		
Familial Mediterranean Fever	249100	1:15
JEWS FROM KURDISTAN		
Thalassemia, alpha	141800	1:8
Thalassemia, beta	141900	1:6
JEWS FROM YEMEN		
Metachromatic leukodystrophy	**250100**	**1:50**
Phenylketonuria	**261600**	**1:35**
Pituitary dwarfism II (Laron)	262500	unknown
Thalassemia, alpha	141800	1:5

However, since the marriage habits among the majority of the Jewish population is by choice, the average couple prefers secondary prevention: to take part in a genetic test only after marriage in order to prevent the future birth of an affected child. But due to the fact that the only way to achieve the purpose of such screening is by interruption of pregnancy, it should be proposed only for tests, which have a potential to justify the interruption of a pregnancy. For this measure to be well accepted there should be a consensus within the target population that the severity of the disease is such that therapeutic abortion may be envisaged. This was one of the main reasons for acceptance of the the Tay-Sachs carrier screening among Ashkenazi Jews which also included the religious community, even if most of the religious members did not use the program for themselves.

In a program aimed at preventing severe genetic diseases or disorders, the criteria of severity should not only refer to the clinical disease but should also include the mutations. The objective of the screening should be to detect those mutations, which have been associated with a severe phenotype, and not towards the carriers of all possible mutations. A demonstrative example is the 5T polymorphism in the CFTR gene, which may cause agenesis of vas deferens [41]. This mutation should not be included in a genetic screening of the general population since the aim of a national screening is not to detect fetuses at risk to be infertile. Another example is the case of non-neuronopathic Gaucher disease, a disease frequent in the Ashkenazi Jewish community. The most frequent mutation [GBA, N370S] is often associated with a mild type of the disease while the second most frequent mutation [GBA, 84GG] is associated with a severe disease. The frequency of a very mild form of the disease, the inability to predict the severity of the phenotype and the existence of an effective treatment for this phenotype are among the arguments not to recommend screening for carriers of Gaucher disease in the Ashkenazi Jewish population. Since the mild mutation [GBA, N370S] is particularly problematic, a possibility is to screen only for carriers of the mutations causing a severe form of the disease. The problem then is that these mutations are rare and their frequency does not justify their inclusion in a screening program.

5. Conclusion

Genetic screening has been offered only to the Ashkenazi Jewish population and relatively recently the screening has been expanded to other Jewish communities. Several screening programs aimed at the Ashkenazi Jews exist in Israel and the USA, offering all a similar core of tests for the frequent severe diseases, such as Tay-Sachs, Cystic fibrosis and Familial dysautonomia [42]. However, many of these programs differ from the other diseases screened. The recommendation of the American College of Obstetricians and Gynecologists (ACOG) committee on Genetics is that the Ashkenazi Jews will be offered screening for Tay-Sachs disease, cystic fibrosis, familial dysautonomia and Canavan disease. The committee also stated that "couples of Ashkenazi Jewish descent may inquire about the availability for other disorders: Mucolipidosis IV, Niemann Pick disease type A, Fanconi anemia group C syndrome and Gaucher disease" [43]. The Israeli Society of Medical Geneticists [43], recently delineated criteria for genetic screening in Israel. According to those criteria, the genetic tests for carrier detection are highly recommended for severe diseases with a frequency higher than 1:15,000. The recommendation is also to discuss with the couples the possibility of screening for other severe diseases with a frequency between 1:15,000 and 1:40,000. The screening for disorders such as Gaucher disease or deafness is not recommended in the reproductive screening context. These criteria should be applied to the entire Jewish population since they are consistent with the international recommendations. They may be easily applied according to the origin of each of the spouses and the frequency of the diseases in the different communities. The decision whether a disease is severe enough to justify screening should be taken according to the existing consensus in the particular community.

References

[1] Goodman R. M., (1979). Genetic disorders among the Jewish people. *The John Hopkins University Press.*

[2] Zlotogora J., Bach G., Munnich M., (2000). *The molecular basis of Mendelian disorders among Jews*. Mol Genet Metab 69:169-180.

[3] Zlotogora J. Mendelian disorders among Jews. www.health.gov.il/genetics.

[4] Kaback M, Lim-Steele J, Dabholkar D, Brown D, Levy N, Zeiger K., (1993). Tay-Sachs disease-carrier screening, prenatal diagnosis, and the molecular era. An international perspective, 1970 to 1993. The International TSD Data Collection Network. *JAMA* 270:2307-2315,

[5] Goldschmidt E, Ronen A, Ronen I., (1960). Changing marriage system in the Jewish communities in Israel. *Ann Hum Genet* 24:191-204.

[6] Cohen T, Vardi-Saliternik R, Friedlander. Consanguinity, intracommunity and intercommunity marriages in a population sample of Israeli Jews. *Ann Hum Biol* 31:38-48, 2004.

[7] Paw BH, Tieu PT, Kaback MM, Lim J, Neufeld EF. Frequency of the three Hex A mutant allele among Jewish and non-Jewish carriers identified in a Tay-Sachs screening program. *Am J Hum Genet* 47: 698-705, 1990.

[8] Kerem E, Kalman YM, Yahav Y, Shoshani Z, Abeliovich D, Szeinberg A, Rivlin J, Blau H, Tal A, Ben-Tur L, Springer C, Augarten A, Godfrey S, Lerer I, Branski D, Friedman M, Kerem B. Highly variable incidence of cystic fibrosis and different mutation distribution among different Jewish ethnic groups in Israel. *Hum Genet* 96: 193-197, 1995.

[9] Anderson SL, Coli R, Daly IW, Kichula A, Rork MJ, Volpi SA, Ekstein J, Rubin BY. Familial dysautonomia is caused by mutations of the IKAP gene. *Am J Hum Genet* 68:753-758, 2001. Epub Jan 22. 2001.

[10] Kaul R, Gao P, Aloya M, Balamurugan K, Petrosky A, Michals K, Matalon R. Canavan disease; mutations among Jewish and non-Jewish patients. *Am J Hum Genet* 55: 34-41, 1994

[11] Eng LL, Desnick RJ, German J, Ellis NA Carrier frequency of the Bloom syndrome blmAsh mutation in the Ashkenazi Jewish population. *Mol Genet Metab* 64:286-290, 1998.

[12] Tamary H, Bar-Yam R, Zemach M, Dgany O, Shalmon L, Yaniv I. The molecular biology of Fanconi anemia. *Isr Med Assoc J* 4:819-823, 2002.

[13] Levran O, Desnick RJ, Schuchman EH. Identification and expression of a common missense mutation (L302P) in the acid sphingomyelinase gene of Ashkenazi Jewish type A Niemann-Pick disease patients *Blood* 80:2081-2087, 1992.

[14] Bargal R, Avidan N, Ben-Asher E, Olender Z, Zeigler M, Frumkin A, Raas-Rothschild A, Glusman G, Lancet D, Bach G. Identification of the gene causing mucolipidosis IV. *Nat Genet* 26:118-121, 2000.

[15] Shaag A, Saada A, Berger I, Mandel H, Joseph A, Feigenbaum A, Elpeleg ON: Molecular basis of lipoamide dehydrogenase deficiency in Ashkenazi Jews. *Am J Med Genet* 82:177-182, 1999.

[16] Ekstein J, Rubin BY, Anderson SL, Weinstein DA, Bach G, Abeliovich D, Webb M, Risch N. Mutation frequencies for glycogen storage disease Ia in the Ashkenazi Jewish population. *Am J Med Genet* 129A:162-164, 2004.

[17] Edelmann L, Wasserstein MP, Kornreich R, Sansaricq C, Snyderman SE, Diaz G. Mapple Syrup Urine Disease: identification and carrier frequency determination of a novel founder mutation in the Ashkenazi Jewish population. Am J Hum Genet 69:863-868, 2001.

[18] Anderson SL, Ekstein J, Donnelly MC, Keefe EM, Toto NR, LeVoci LA, Rubin BY. Nemaline myopathy in the Ashkenazi Jewish population is caused by a deletion in the nebulin gene. Hum Genet 115:185-190, 2004.

[19] Ben-Yosef T, Ness SL, Madeo AC, Bar-Lev A, Wolfman JH, Ahmed ZM, Desnick RJ, Willner JP, Avraham KB, Ostrer H, Oddoux C, Griffith AJ, Friedman TB. A mutation of PCDH15 among Ashkenazi Jews with the type 1 Usher syndrome. *N Engl J Med* 348:1664-1670, 2003.

[20] Beutler E, Gelbart T, Kuhl W, Zimran A, West C. Mutations in Jewish patients with Gaucher disease. *Blood* 79:1662-1666, 1992.

[21] Morell RJ, Kim HJ, Hood LJ et al. Mutations in the connexin 26 gene (GJB2) among Ashknenazi Jews with non syndromic recessive deafness. *New Engl J Med* 339:1500-1505, 1998.

[22] Lerer I, Sagi M, Ben-Neriah Z, Wang T, Levi H, Abeliovich D. A deletion mutation in GJB6 cooperating with GJB2 mutation in trans in non-syndromic deafness: a novel founder mutation in Ashkenazi Jews. *Hum Mut. Mutations in brief* 458 on line, 2001.

[23] Gilad S, Bar-Shira A, Harnik R, Shkedy D, Ziv Y, Khosravi R, Brown K, Vanagaite L, Xu G, Frydman M, Lavin MF, Hill D, Tagle DA, Shiloh Y. Ataxia telangiectasia: founder effect among North African Jews. *Hum Mol Genet* 5: 2033-2037, 1996.

[24] Parvari R, Moses S, Shen J, Hershkovitz E, Lerner A, Chen YT. A single base deletion in the 3' coding region of glycogen debranching enzyme is prevalent in glycogen storage disease type IIIA in a population of North African Jewish patients. *Eur J Genet* 5:266-270, 1997.

[25] Kalatzis V, Cohen-Solal L, Cordier B, Frishberg Y, Kemper M, Nuutinen EM, Legrand E, Cochat P, Antignac C. Identification of 14 novel *CTNS* mutations and characterization of seven splice site mutations associated with cystinosis. *Human Mutation* 20:439-446, 2002.

[26] Rosler A, Lieberman E , Cohen T. High frequency of congenital adrenal hyperplasia (11 ß hydroxylase deficiency) among Jews from Morocco. *Amer J Med Genet* 42 :827-834, 1992.

[27] Leitersdorf E, Reshef A, Meiner V, Levitzki R, Schwartz SP, Dann EJ, Berkman N, Cali JJ, Klapholz L, Berginer VM. Frameshift and splice junction mutations in the sterol 27-hydroxylase gene cause cerebrotendinous xanthomatosis in Jews of Moroccan origin. *J Clin Invest* 91:2488-2493, 1993.

[28] Kaufman M, Grinshpun-Cohen J, Karpati M, Peleg L, Goldman B, Akstein E, Adam A, Navon R. Tay Sachs disease and HEXA mutations among Morrocan Jews. *Hum Mut* 10:295-300, 1997.

[29] Ben-Zeev B, Levy-Nissenbaum E Lahat H, Anikster Y, Shinar Y, Brand N, Gross-Tzur V, MacGregor D, Sidi R, Kleta R, Frydman M, Pras E. Megalencephalic Leukoencephalopathy with Subcorical Cysts; a Founder Effect in Israeli Patients and a Higher than expected Carrier Rate Among Libyan Jews. *Hum Genet Hum Genet* 111:214-218, 2002. Epub 2002 Jul 16, 2002.

[30] McNally EM, Ly CT, Rosenmann H, Mitrani Rosenbaum S, Jiang W, Anderson LVB, Soffer D, Argov Z. Splicing mutation in dysferlin produces limb girdle muscular dystrophy with inflammation. *Am J Med Genet* 91:305-312, 2000.

[31] Ohno K, Sadeh M, Blatt I, Brengman JM, Engel AG.E-box mutations in the RAPSN promoter region in eight cases with congenital myasthenic syndrome. *Hum Mol Genet* 12:739-748, 2003.

[32] Anikser Y, Kleta R, Shaag A, Gahl AW, Elpeleg O. Type III 3 Methylglutaconic aciduria (optic atrophy plus syndrome or Costeff optic atrophy syndrome): identification of the OPA3 gene and its founder mutation in Iraqi Jews. *Am J Hum Genet* 69:1218-1224, 2001.

[33] Rosenberg N, Yatuv R, Orion Y, Zivelin A, Pertz H, Seligsohn U. Glanzmann thrombasthenia caused by an 11.2-kb deletion in the glycoprotein IIIa (beta3) is a second mutation in Iraqi Jews that stemmed from a distinct founder. *Blood* 15: 3654-3662, 1997.

[34] Eisenberg I, Avidan N, Potikha , Hochner H, Chen M, Olender M, Olender T, Barash M, Shemesh , Sadeh M, Grabov-Nardini G, Shmilevich I, Friedmann A, Karpati G, Bradley WG, Baumbach L, Lancet D Ben Asher E, Beckmann J, Argov Z, Mitrani-Rosenbaum S. The UDP-Nacetylglucosamine 2-epimerase /N-acelylmannosamine kinase gene is mutated in recessive hereditary inclusion body myopathy. *Nature Genet* 29:83-87, 2001.

[35] Bjorses P, Aaltonen J, Vikman A, Perheentupa J, Ben-Zion G, Chimuello G, Dahl N, Heideman P. Hoorweg-Nijman JJG., Mathivon L, Mullis PE, Pohl M, Rizen M, Romeo G, Shapiro MS, Smith CS, Solyom J, Zlotogora J, Peltonen L. Genetic homogeneity of autoimmune polyglandular syndrome type I. *Am J Hum Genet* 59:879-886, 1996.

[36] Rund D, Filon D, Doewling CE, Rachmilewitz E, Kazazian HH Jr, Oppenheim A. Diversity of molecular lesions causing beta thalassemia in Israel Jewish ethnic groups. in New perspectives in genetic markers and diseases among the Jewish people. Ed Adam A, Bonne Tamir B 1991. Oxford Press.

[37] Avigad S, Cohen BE, Bauer S, Schwartz G, Frydman M, Woo SLC, Niny Y, Shiloh Y. A single origin of phenylketonuria in Yemenite Jews. *Nature* 344:168-170, 1990.

[38] Zlotogora J, Bach G, Bosenberg C, Barak Y, von Figura K, Gieselmann V. Molecular basis of late infantile metachromatic leukodystrophy in the Habbanite Jews. *Hum Mutation* 5:137-143, 1995.

[39] Wilson JMG, Jungner G. Principles and practice of screening for diseases. *Geneva WHO* 1968.

[40] Broide E, Zeigler M, Eckstein J, Bach G. Screening for carriers of Tay-Sachs disease in the ultra orthodox Ashkenazi Jewish community in Israel. *Am J Med Genet* 47:213-215, 1993.

[41] American College of Medical Genetics available at: http://www.acmg.net/resources/policy-list.asp.

[42] Leib JR, Gollust SE, ChandrosHull S, Wilfond B. Carrier screening panels for Ashkenazi Jews: is more better? *Amer J Med Genet* 7:185-190, 2005.

[43] ACOG Committee on Genetics. ACOG committee opinion. Number 298, August 2004. Prenatal and preconceptional carrier screening for genetic diseases in individuals of Eastern European Jewish descent. *Obstet Gynecol.* 104:425-428, 2004.

[44] Israeli Society for Medical geneticists. 2003.

In: Focus on Genetic Screening Research
Editor: Sandra R. Pupecki, pp. 73-93
ISBN 1-60021-229-8

Chapter IV

Polymorphisms of Human Platelet Receptor Genes: Clinical Relevance and Diagnostic Approaches

Peter Bugert**, *Marion Vosberg, Andrea B. Lese and Harald Klüter
Institute of Transfusion Medicine and Immunology, University of Heidelberg,
Faculty of Clinical Medicine, Mannheim, Germany

Abstract

Human blood platelets represent cells with a variety of important functions in hemostatic, immunological and inflammatory processes. Glycoproteins (GP) and other receptors are expressed on the cell surface and are involved in platelet adhesion to matrix proteins or cells. This review is intended to give a summary of the current knowledge about the clinical relevance of single nucleotide polymorphisms (SNPs) and mutations in platelet receptor genes. Some SNPs represent the genetic basis of the human platelet alloantigen (HPA) systems. The gene variants are considered in the context of platelet disfunction as well as in relation to the pathogenesis of vascular diseases and thromboembolic events. Other polymorphic adhesion receptors expressed on platelets are, for example, P-selectin, PSGL-1 and PECAM-1. Results from case/control studies including patients with vascular diseases or thromboembolic events are summarized. The review also considers the genetic typing techniques currently available for the platelet receptor polymorphisms. A variety of techniques for genotyping platelet receptor polymorphisms, mostly HPA SNPs, have been developed and some of them are nowadays routinely used in specialized clinical diagnosis. Prenatal HPA-typing mostly done on fetal DNA obtained by amniocentesis or umbilical cord vein puncture represents a relevant aspect in the early diagnosis of neonatal alloimmune thrombocytopenia (NAIT). The use of fetal DNA circulating in the maternal plasma is a challenge for future developments. The HPA-

* Correspondence address: Peter Bugert, Ph.D. Institute of Transfusion Medicine and Immunology; German Red Cross Blood Service of Baden-Württemberg – Hessen gGmbH; Friedrich-Ebert-Straße 107; D-68167 Mannheim, Germany; Phone number: +49-621-3706-8122, Fax number: +49-621-3706-876; e-Mail: p.bugert@blutspende.de

typing approaches are predominantly performed as individual testing and are only practicable for sample numbers in a low throughput range. Other techniques are currently developed on the basis of microarray platforms and will allow a complex typing of large numbers (>100) of reasonable SNPs in a single analysis. In future, these techniques may be used for both, the low throughput individual testing in clinical diagnosis and the high throughput screening of blood donors for certain constellations of blood cell antigens.

Introduction

Blood platelets are anucleated cell fragments derived from megakaryocytes in bone marrow and represent the most important cellular component in hemostasis. They also play an important part in maintaining the integrity of the blood vessel endothelium, in wound healing and in tissue regeneration. A large variety of growth factors, cytokines and other effector molecules can be found in the storage granules of platelets and are released upon activation. Platelets may therefore be regarded as immunological and inflammatory effector cells.

The different platelet functions are based on their binding to extracellular matrix proteins or their adhesion to cells, such as blood vessel endothelium or leukocytes. The specific interaction is mediated by receptor molecules expressed on the platelet outer surface. Among the different kinds of receptors the glycoproteins (GP) represent the most abundant class of molecules specifically expressed on platelets. Different kinds of polymorphisms (allele frequency >1 %), most of them 'single nucleotide polymorphisms' (SNPs), or mutations (allele frequency <1 %) were described for the GPs which presumably alter their antigenicity, expression level or function. The human platelet antigen (HPA) nomenclature system was introduced in 1990 and all alloantigens described so far were localized on GPs, except HPA-15 caused by a CD109 polymorphism (reviewed in Metcalfe *et al.,* 2003). Other receptors can be subsumed as adhesion molecules and polymorphisms, for example P-selectin, PSGL-1 (P-selectin glycoprotein ligand 1) or PECAM-1 (platelet/endothelial cell adhesion molecule 1) which were described as well (reviewed in Blankenberg *et al.,* 2003).

This article reviews polymorphisms and mutations of platelet receptor genes, especially those affecting the protein sequence and function. We focus on the clinical relevance of these gene variants and on the genotyping methods feasible for routine diagnosis.

1. Platelet Glycoprotein Polymorphisms and Mutations

1.1. GPIIb/IIIa Complex

The heterodimeric GPIIb/IIIa complex, also named integrin $\alpha_{IIb}\beta_3$, is the numerically predominant GP on the platelet surface and mediates binding to fibrinogen and von Willebrand factor (vWF). Eleven polymorphisms causing alloantigens have been described for this GP complex so far (Table 1). The HPA-3a/b and HPA-9bw antigens are present on GPIIb and the genetic basis of both antigens involves SNPs leading to a single amino acid change. For GPIIIa 9 alloantigens (HPA-1a/b, -4a/b, -6bw, -7bw, -8bw, -10bw, -11bw, -14bw and -16bw) have been identified so far and SNPs are the genetic basis of most of them. Only HPA-14bw is an exception in that it results from the deletion of an entire codon and the subsequent deletion of amino acid 611 from the mature protein. The bw-antigens are serologically defined and alloantibodies against the thetical, but not the antithetical antigens, have been observed in single clinical cases. Thus, the genetic basis of bw-antigens rather represents 'private' mutations than polymorphisms. In addition, mutations in the GPIIb/IIIa complex are the most common cause of Glanzmann thrombasthenia (GT), an inherited bleeding disorder (reviewed in Nurden and Nurden, 2002). The mutations either show quantitative effects on the level of GPIIb/IIIa surface expression or functional effects on activation or ligand binding of the receptor. Table 1 only lists the missense and nonsense mutations. Further details about deletions, insertions and intervening sequences have been reviewed in a different publication (Nurden and Nurden, 2002).

1.2. GPIb/IX/V Complex

The GPIb/IX/V complex contains four different subunits (GPIbα, GPIbß, GPV, GPIX) and is a receptor with a variety of ligand specificities (reviewed in Lopez and Berndt, 2002). GPIb/IX/V binding to subendothelial von Willebrand factor (VWF), thrombin and certain coagulation factors thus highlighting its role of in hemostasis and thrombosis. The receptor also mediates binding of platelets to the endothelium via P-selectin interaction and binding to leukocytes via Mac-1 interaction. GPIb/IX/V deficiency is the cause of the Bernard-Soulier syndrome (BSS), a bleeding disorder with autosomal recessive inheritance (reviewed in Lopez *et al.,* 1998). Various mutations of GPIbα, GPIbß and GPIX have been identified in BSS patients (Table 2) and affect expression levels of the GP complex or ligand binding activity (and Nurden and Nurden, 2002). Platelet-type von Willebrand disease (plt-VWD) is another bleeding disorder caused by mutations affecting the GPIb/IX/V complex. The mutations described so far were all located in a loop region of GPIbα formed by a disulfide bond between Cys209 and Cys248 (Miller *et al.*, 1991; Russell and Roth, 1993; Matsubara *et al.*, 2003). The GPIbα gene also contains a variable number of tandem repeats (VNTR) polymorphism encoding 13 amino acids per repeat unit. The rare A-allele contains 4 repeat units, whereas the most common C-allele contains 2 repeats. Compared to the strong association of GPIbα mutations with BSS or plt-VWD, none of the VNTR alleles were related to hemostatic or thrombotic complications. Some clinical studies provide support for the VNTR B-allele as a potential cardiovascular risk factor (reviewed in Yee and Bray, 2004). In addition to the disease-related mutations of the GPIb/IX/V complex, two alloantigens have been described. A SNP in the GPIbα gene is the genetic basis of the biallelic HPA-2a/b alloantigen system and a mutation of GPIbß leading to the amino acid change 15Gly>Glu causes the private antigen HPA-12bw.

Table 1. SNPs and mutations of the GPIIb/IIIa complex

Glycoprotein	Integrin	CD	Nucleotide change[1]	Amino acid change[2]	Phenotype[3]	Allele frequency[4]
GPIIb	α_{IIb}	CD41	520T>C	143Tyr>His	GT	
			526C>G	145Pro>Ala	GT	
			528C>T	145Pro>Leu	GT	
			620C>T	176Thr>Ile	GT	
			641T>C	183Leu>Pro	GT	
			818G>A	242Gly>Asp	GT	
			1063G>A	324Glu>Lys	GT	
			1073G>A	327Arg>His	GT	
			1346G>A	418Gly>Asp	GT	
			1750C>T	553Arg>stop	GT	
			1787T>C	565Ile>Thr	GT	
			2113T>C	674Cys>Arg	GT	
			2333A>C	747Gln>Pro	GT	
			2602G>A	837Val>Met	HPA-9bw	0.97/0.03
			2609G>A	839Trp>stop	GT	
			2621T>G	843Ile>Ser	HPA-3a/b	0.61/0.39
			2929C>T	946Arg>stop	GT	
			2941C>T	950Gln>stop	GT	
			3077G>A	995Arg>Gln	GT	
GPIIIa	β_3	CD61	176T>C	33Leu>Pro	HPA-1a/b	0.85/0.15
			263G>A	62Arg>Gln	HPA-10bw	
			356G>A	93Arg>Gln	HPA-1a	
			428T>G	117Leu>Trp	disrupture	
			433G>T	119Asp>Tyr	GT	
			433G>A	119Asp>Asn	GT	
			497C>T	140Thr>Ile	GT	
			506G>A	143Arg>Gln	HPA-16bw	
			563C>T	162Ser>Leu	HPA-4a/b	
			665T>C	196Leu>Pro	GT	
			718C>T	214Arg>Trp	GT	
			719G>A	214Arg>Gln	GT	
			725G>A	216Arg>Gln	GT	
			727G>A	217Asp>Asn	GT	
			863T>C	262Leu>Pro	GT	
			917A>C	280His>Pro	GT	
			1219G>A	374Cys>Tyr	GT	
			1297C>G	407Pro>Ala	GT	
			1448G>A	457Cys>Tyr	HPA-7bw	
			1544G>A	489Arg>Gln	GT	
			1702T>C	542Cys>Arg	HPA-6bw	
			1756T>C	560Cys>Arg	GT	
			1757G>T	560Cys>Phe	GT	
			1813G>A	579Gly>Ser	GT	
			1909-11del	611Lys>del	GT	
			1976G>A	659Arg>His	HPA-14bw	
			1984C>T	636Arg>Cys	HPA-11bw	
			2248C>T	724Arg>stop	HPA-8bw	
			2332T>C	752Ser>Pro	GT	
					GT	

[1]numbers refer to mRNA reference sequences and the A in the ATG start codon is defined as nucleotide position 1 (GPIIb, NM_000419; GPIIIa, NM_000212); frameshift mutations, deletions (except HPA-14) and insertions are not listed;

[2]numbers refer to the amino acid sequence of the mature proteins;

[3]phenotype refers to the effect of the amino acid change on the protein structure, such as the immunologically relevant HPA systems, or on the protein function, such as mutations leading to Glanzmann thrombasthenia (GT);

[4]numbers represent the allele frequency of diallelic SNPs in Caucasian populations.

Table 2. Polymorphisms and mutations of the GPIb/IX/V complex and other GP

Glycoprotein	Integrin	CD	Nucleotide change[1]	Amino acid change[2]	Phenotype[3]	Allele frequency[4]
GPIa	α_2	CD49b	1600G>A	505Glu>Lys	HPA-5a/b	0.87/0.13
			2483C>T	799Thr>Met	HPA-13bw	
GPIbα		CD42b	217C>T	57Leu>Phe	BSS	
			241T>C	65Cys>Arg	BSS	
			256C>T	70Leu>Phe		(0.96/0.04)
			424C>T	126Leu>Phe	BSS	
			434T>C	129Leu>Pro	BSS	
			482C>T	145Thr>Met	HPA-2a/b	0.93/0.07
			515C>T	156Ala>Val	BSS	
			673T>A	209Cys>Ser	BSS	
			745G>A	233Gly>Ser	plt-VWD	
			746G>T	233Gly>Val	plt-VWD	
			763A>G	239Met>Val	plt-VWD	
			1077G>A	343Trp>stop	BSS	
			1379C>A	444Ser>stop	BSS	
			1542G>A	498Thr>stop	BSS	
			1243-1398 (39 bp)$_4$	399-450 (13 aa)$_4$	VNTR-allele A	<0.01
			1243-1359 (39 bp)$_3$	399-437 (13 aa)$_3$	VNTR-allele B	0.08
			1243-1320 (39 bp)$_2$	399-425 (13 aa)$_2$	VNTR-allele C	0.81
			1243-1281 (39 bp)$_1$	399-411 (13 aa)$_1$	VNTR-allele D	0.11
			del1243-1281	del399-411	VNTR-allele E	<0.01
GPIbß		CD42c	119G>A	15Gly>Glu	HPA-12bw	
			138G>A	21Trp>stop	BSS	
			266A>C	64Asn>Thr	BSS	
			296C>G	74Pro>Arg	BSS	
			338A>G	88Tyr>Cys	BSS	
			397G>C	108Ala>Pro	BSS	
GPIIa	β_1	CD29	623C>T	188Thr>Ile	SCC?	
GPVI			655T>C	199Ser>Pro	GPVI-a/b	0.85/0.13
			709A>C	217Lys>Glu	(reduced function	
			745A>G	229Thr>Ala	and expression for	
			950A>T	297Glu>Leu	b-allele)	
			964C>A	302His>Asn		
GPIX		CD42a	110A>G	21Asp>Gly	BSS	
			167T>C	40Leu>Pro	BSS	
			179A>G	45Asn>Ser	BSS	
			212T>C	55Phe>Ser	BSS	
			266G>A	73Cys>Tyr	BSS	
			338G>A	97Cys>Tyr	BSS	
			426G>A	126Trp>stop	BSS	
			463G>A	139Ala>Thr	BSS	
		CD109	2108C>A	682Ser>Tyr	HPA-15a/b	0.51/0.49

[1]numbers refer to mRNA reference sequences and the A in the ATG start codon is defined as nucleotide position 1 (GPIa, NM_002203; GPIbα, NM_00173; GPIbß, NM_00407; GPIIa, NM_002211; GPIV, NM_000072; GPVI, NM_016363; GPIX, NM_000174; CD109, NM_133493);

[2]numbers refer to the amino acid sequence of the mature proteins;

[3]phenotype refers to the effect of the amino acid change on the protein structure, such as the immunologically relevant HPA systems, or on the protein function, such as mutations leading to Bernard-Soulier syndrome (BSS) or platelet-type von Willebrand disease (plt-VWD); a GPIIa mutation has been reported from a squamous cell carcinoma (SCC) cell line, but its role in tumor development remains unclear;

[4]numbers represent the allele frequency of polymorphisms in Caucasian populations (or Japanese population).

1.3. Other GP Complexes

The GPIa/IIa complex, also known as integrin $\alpha_2\beta_1$, is one of the platelet receptors for binding to collagen. The alloantigens HPA-5a/b and HPA-13bw are located on GPIa, whereas, no alloimmunization against GPIIa has been described so far (Table 2). A GPIIa mutation leading to the amino acid substitution 188Thr>Ile was described in a cell line derived from a squamous cell carcinoma. Whether the mutation is involved in tumor development remains unclear, but it results in constitutive activation of ligand binding (Evans *et al.*, 2003).

GPVI is the major signaling receptor for collagen on platelets that forms a complex with the Fc receptor γ chain. Exon re-sequencing of the GPVI coding region revealed a high resolution SNP map by which 2 common alleles, GPVI-a and -b, could be described (Croft *et al.*, 2001; Joutsi-Korhonen *et al.*, 2003). The two alleles differ by 5 amino acid replacements: 199Ser>Pro, 217Lys>Glu, 229Thr>Ala, 297Glu>Leu, 302His>Asn. Ligand binding was significantly lower for the b-allele and platelets homozygous for this variant tended to be less able to form a thrombus on collagen in flowing blood (Joutsi-Korhonen *et al.*, 2003). So far, no alloimmunization against the GPVI variants has been reported.

The phosphatidylinositol- (GPI) anchored platelet glycoprotein CD109 carries the biallelic HPA-15a/b antigen system (Schuh *et al.*, 2002). The genetic basis was characterized as a missense SNP at position 2108 of the coding region, resulting in a Ser>Tyr substitution at position 682 of the mature protein.

1.4. HPA in Clinical Platelet Immunology

Alloimmunization against platelet antigens is involved in clinical complications such as neonatal alloimmune thrombocytopenia (NAIT), posttransfusion purpura (PTP) or platelet transfusion refractoriness (PTR) (reviewed in Metcalfe, 2004). In NAIT, maternal IgG against fetal platelet alloantigens cross the placenta and cause immune destruction of platelets *in utero*. The severity of the disease ranges from asymptomatic low platelet counts to intracranial hemorrhage and death. Approximately 80-90 % of cases are due to alloimmunization of the HPA-1b homozygous mother against the paternal inherited HPA-1a allele of the fetus. Antibodies against HPA-5b can be observed in 5-15 % of the cases. The remaining NAIT cases are caused by the other biallelic HPA systems or the 12 private antigens (-bw) described up until now. Recently, the relevance of alloimmunization against HPA-15 antigens has been estimated (Ertel *et al.*, 2005). Two families with NAIT were reported and antibodies against HPA-15b were identified in addition to HLA-antibodies. Anti-HPA-15 antibodies showed a higher incidence in polytransfused patients compared to NAIT cases (3.0 % vs. 0.2 %).

A rare clinical complication, PTP, can be observed after transfusion of platelets to patients who had been previously exposed to allogeneic platelets, either through pregnancy or blood transfusion. In most of the cases a high-titre anitbody against HPA-1a can be observed in the patient's serum. The pathomechanism due to which the transfused as well as the own platelets (not carrying the alloantigen) are simultaneously desrupted is unknown.

PTR often occurs in patients after long-term platelet transfusion due to the development of HLA antibodies (reviewed in Novotny, 1999). In addition, HPA antibodies can be observed and the low frequency alleles HPA-1b, -2b and -5b seem to be commonly involved (Kurz *et al.*, 1996).

1.5. HPA in Transplantation Immunology

The HPA systems have been considered to be minor histocompatibility antigens (mHA) in clinical bone marrow, peripheral blood progenitor cells or renal transplantation. Several studies on patients who received bone marrow derived grafts from an HLA-identical sibling, did not provide evidence that HPA systems should be considered as mHA (Balduini *et al.*, 1999; Balduini *et al.*, 2001; Rozman *et al.*, 2003). The occurence of acute graft-versus-host disease (aGVHD) could be associated with HPA-3 incompatibility in some patients. In a Japanese study including 715 unrelated bone marrow transplantations, the HPA-5 system was found to have a significant effect on the disease-free survival rate of recipients but not on the development of aGVHD (Juji *et al.*, 1999).

The role of HPA systems as mHA in the transplantation of peripheral blood derived stem cells has been investigated as well (Garcia-Malo *et al.*, 2004; Leitner *et al.*, 2005). In conclusion, the HPA match does not affect the outcome of transplantation or the occurence of immune complications.

The acute vascular rejection (AVR) of a kidney graft is an immunologically mediated complication in transplantation. Alloantigenic epitopes expressed by the vascular endothelium were considered as potential targets. Because integrins β3, α2 and GPIb are also expressed by endothelial cells, the HPA-systems (HPA-1, -2 and -5) carried by these receptors were regarded as candidates involved in AVR development (Kekomäki *et al.*, 2001). The results suggested that HPA-5b represents a mHA associated with the development of AVR.

1.6. HPA in Thromboembolism

Platelet GP complexes play an important role in hemostasis and platelet adhesion. Therefore, polymorphisms that affect the expression or function of the receptors could influence the risk for development or outcome of thromboembolic disease. The biallelic HPA SNPs belong to the candidates for case-control studies including patients with vascular diseases or thromboembolic complications. They have been widely studied in various populations and patient cohorts with coronary heart disease (CHD), myocardial infarction (MI) or stroke (reviewed in Bussel *et al.*, 2000). In summary, most results were conflicting each other and could not be reproduced in other cohorts of the same clinical entity. HPA alleles are, therefore, unlikely to represent genetic risk factors for the development of vascular or thromboembolic disease (reviewed in Meisel *et al.*, 2004). Zotz *et al.* (1998) reported that HPA-1b is a risk factor for platelet thrombogenicity in patients with preexisting CHD. The platelet antigen was associated with a higher risk of MI in these patients. In addition, the HPA-1b prevalence was significantly higher among patients with postoperative complications after coronary bypass surgery (Zotz *et al.*, 2000). Thus, the HPA-1b allele may be regarded as a risk factor for a prothrombotic state from which certain clinical complications can arise.

2. Adhesion Receptor Polymorphisms and their Clinical Relevance

Adhesion of platelets to other cells (leukocytes or vascular endothelium) is not only mediated by GP complexes but also by adhesion receptors. Selectins represent important molecules involved in early steps of cell adhesion, such as the rolling of circulating blood cells on the endothelium (reviewed in Vestweber and Blanks, 1999). Subsequently, integrins, molecules of the immunoglobulin superfamily (PECAM-1) and others allow the firm attachment to and transmigration through the vessel wall. Platelet adhesion to leukocytes is predominantly mediated via the interaction of P-selectin (platelets) and PSGL-1 (leukocytes) (reviewed in Yang *et al.*, 1999) or JAM-3 (platelets) and MAC-1 (leukocytes) (Santoso *et al.*, 2002). Adhesion of circulating cells to the arterial surface is among the first detectable events in atherogenesis leading to major complications (myocardial infarction, ischemic stroke and peripheral occlusion). The involved adhesion receptors belong to the first choice candidate genes for polymorphism screening and epigenetic case/control studies.

2.1. P-Selectin

P-selectin, expressed on stimulated endothelium and activated platelets, is one of the molecules that has been extensively studied for polymorphisms and its association with disease (reviewed in Blankenberg *et al.*, 2003). Exon re-sequencing of the entire coding region revealed 5 SNPs leading to amino acid replacements (Table 3) (Herrmann *et al.*, 1998; Bugert *et al.*, 2004). The allele with proline at position 715 (Pro715) showed a protective effect against the development of MI since its prevalance was higher in control individuals (Herrmann *et al.*, 1998). Haplotype analysis confirmed the protective effect of the Pro715 allele but additionally revealed that the presence of two asparagine codons at sites S290N and N562D was associated with a higher risk of MI (Tregouet *et al.*, 2002). In contrast, among patients with CHD, the Pro715 allele showed a higher prevalence in individuals with hypercolesterolemia, a classical risk factor for MI (Bugert *et al.*, 2004). Although the relevance of P-selectin variants as genetic risk factor for the development of thromboembolic complications remains uncertain, it may reach greater importance as a disease marker in future (reviewed in Kappelmayer *et al.*, 2004).

Table 3. Polymorphisms and mutations of platelet adhesion receptors

Receptor	CD	Nucleotide change[1]	Amino acid change[2]	Phenotype[3]	Allele frequency[4]
P-selectin	CD62P	625G>A	168Val>Met		0.93/0.07
		992G>A	290Ser>Asn		0.83/0.17
		1807G>A	562Asn>Asp		0.59/0.41
		1918T>G	599Leu>Val		0.11/0.89
		2266A>C	715Thr>Pro		0.86/0.14
PSGL-1	CD162	186G>A	62Met>Ile		0.95/0.05
		818C>T	273Ser>Phe		0.99/0.01
		820A>G	274Met>Val		0.99/0.01
		352-831 (30 bp)$_{16}$	118-277 (10 aa)$_{16}$	VNTR-allele A	0.82
		del382-411 (30 bp)$_{15}$	del128-137 (10 aa)$_{15}$	VNTR-allele B	0.17
		del592-651 (30 bp)$_{14}$	del198-217 (10 aa)$_{14}$	VNTR-allele C	0.01
PECAM1	CD31	373G>C	125Val>Leu		(0.55/0.45)
		1688A>G	563Asn>Ser		(0.50/0.50)
		2008G>A	670Gly>Arg		(0.50/0.50)
CD40L	CD154	655A>G	219Gly>Arg		0.99/0.01
		(see CD40L database at: http://bioinf.uta.fi/CD40Lbase/index.html)		XHIGM	

[1]numbers refer to mRNA or DNA reference sequences and the A in the ATG start codon is defined as nucleotide position 1 (P-selectin, NM_003005; PSGL-1, U25956; PECAM1, NM_000442; CD40L, NM_000074);

[2]numbers refer to the amino acid sequence of the mature proteins;

[3]in the phenotype column gives designation of alleles or refers to the effect of the amino acid change on the protein function, such as mutations leading to X-linked hyper IgM syndrome (XHIGM);

[4]numbers represent the allele frequency of diallelic SNPs in Caucasian populations (or Japanese population).

2.2. PSGL-1

The most important ligand of P-selectin is PSGL-1. The gene structure is very similar to the GPIbα gene: 1. the entire coding region is encompassed by one exon; 2. both genes encode a mucin-like stalk; 3. the stalk consists of tandem repetition of short amino acid stretches; 4. the number of repeat units is variable (VNTR, variable number of repeat units) and is the basis for the length polymorphism of the proteins (Lopez *et al.,* 1992; Afshar-Kharghan *et al.*, 2001). For the PSGL-1 VNTR the experimental and clinical data revealed a correlation of protein length with binding characteristics and ischemic cerebrovascular disease (CVD). The shorter protein variants (alleles B and C) showed a significantly lower binding capacity and these alleles were associated with a lower risk of developing CVD (Lozano *et al.,* 2001). When assessing the polymorphism in a small cohort of 281 CHD patients and 397 healthy controls, we found a significantly higher prevalence of the shorter alleles in patients than in controls (5.3 % vs. 1.5 %). However, this result could not be confirmed in a larger study including 2,578 CHD patients, 731 patients without CHD and 1,084 blood donor controls (Bugert *et al.,* 2003a). Also after subgrouping of CHD patients regarding occurrence of MI (906 with MI vs. 1,304 without MI) we could not observe a difference in prevalence of PSGL-1 alleles or genotypes. In contrast, Roldan *et al.* (2004) described a lower prevalence of the shorter alleles in a case/control study including 219 patients suffered from premature MI (age ≤45 years) and 594 controls. Generally, case/control studies should be adequate regarding the statistical power especially to detect smaller differences (≤5 % in genotype frequencies). This problem is demonstrated by the power calculation (according to Fleiss *et al.*, 1980) for our two CHD studies to detect the difference of 1.5 % vs. 5.3 % genotype frequency (Table 4). If the difference is slightly smaller (2 % vs. 4 %) the statistical power of the small cohort dramatically decreases from 79 % to 35 % (Bugert *et al.*, 2003a).

Table 4. Power calculation for two case/control studies based on the sample sizes (from Bugert *et al.*, 2003a)

Assumed genotype frequency (control / case group)	Power of study 1 for 397 controls / 281 cases	Power of study 2 for 1,084 controls / 2,578 cases
2 % / 4 %	35 %	91 %
1.5 % / 5.3 %	79 %	99.9 %

2.3. Other Adhesion Receptors

PECAM-1 (CD31) is expressed on the cell surface of endothelial cells, platelets and leukocytes. It contains six Ig-like domains, a short transmembrane domain and a cytoplasmic tail. The missense SNPs reported for PECAM-1 reside in different parts of the protein: 125Val>Leu in the first Ig-like domain; 563Asn>Ser in the sixth Ig-like domain; 670Gly>Arg in the cytoplasmic domain involved in signal transduction (reviewed in Newman, 1997). In a Japanese population of 136 MI patients and 235 healthy controls the linked alleles 125Leu, 563Ser and 670Arg were significantly increased in patients compared with controls (Sasaoka *et al.*, 2001). Regarding the genotype frequencies, the homozygous forms 563Ser/Ser and 670Arg/Arg were also significantly increased in patients. The allele and genotype frequencies of the 125Val>Leu polymorphism were similar in a German case-only study including 2500 CHD patients (Gardemann *et al.*, 2000). Subgrouping of the patients regarding occurrence of MI (1,170 patients with MI vs. 1,330 patients without MI) also failed to detect any significant difference. Thus, there is no evidence for the 125Val>Leu polymorphism of PECAM-1 as a risk factor of CHD or MI. However, this as well as the 563Asn>Ser SNP may serve as mHA in bone marrow transplantation because it was considered a risk factor for aGVHD (Balduini *et al.*, 1999; Balduini *et al.*, 2001). The expression level of membrane-bound and soluble PECAM-1 was found to be elevated in MI patients (reviewed in Samara and Grubel, 2003).

Platelets express two different protease-activated receptors, PAR-1 and PAR-4, both binding to thrombin. To best of our knowledge, polymorphisms have been described only for the non-coding regions of PAR-1 (Arnaud *et al.*, 2000a; Park *et al.*, 2000). None of the polymorphisms were found in association with CHD or MI (Park *et al.*, 2000; Arnaud *et al.*, 2000b). In a French study including 250 patients with venous thrombosis and 1,214 healthy controls the –506I/D polymorphism (5'-untranslated region) revealed a protective effect of the –506I-allele in males (Arnaud *et al.*, 2000a).

CD40L is expressed on activated T-cells and platelets. The gene was mapped to the long arm of the X-chromosome and mutations cause the X-linked hyper IgM syndrome (XHIGM or X-HIM), a rare immune deficiency disorder characterized by low or absent serum levels of IgG, IgA and IgE (reviewed in Fuleihan, 2001; Etzioni and Ochs, 2004). Elevated levels of soluble CD40L can be found in patients with acute coronary syndromes or stroke and is therefore discussed as a potential biomarker (reviewed in Apple *et al.*, 2005). A polymorphic dinucleotide repeat in the 3'-untranslated region of CD40L could be associated with protein expression levels and may serve as a genetic marker for CD40L expression (Citores *et al.*, 2004). In addition, a naturally occurring SNP in CD40L was found (Barnhart *et al.*, 2000), nevertheless its association with vascular or thromboembolic disease remains unclear.

3. Genotyping Techniques and their Diagnostic Application

The majority of current methods for genotyping of SNPs or length polymorphisms are based upon PCR. The PCR technique is well established in many laboratories responsible for routine diagnosis. Usually, genomic DNA, the predominant target of PCR investigation, is isolated from anticoagulated blood samples but may also be obtained from other biological samples (amniocytes, buccal swaps, etc.). Discrimination between SNP alleles may be achieved during PCR amplification by using sequence or allele specific primers (PCR-SSP or PCR-ASP) or by using fluorescently labelled allele specific probes in TaqMan™ 5'-nuclease assays (Lee *et al.*, 1993; reviewed in Hurd *et al.*, 2002). In other approaches post-PCR procedures are performed to achieve allelic discrimination, such as PCR with sequence or allele specific oligonucleotides (PCR-SSO or PCR-ASO), PCR followed by restriction enzyme digest (PCR-RFLP, -restriction fragment length polymorphism) and others.

3.1. Low Throughput Individual Testing

The biallelic HPA-Systems (HPA-1, -2, -3, -5 and -15) represent the clinically most relevant platelet SNPs. Typing of the platelet antigens and antibodies is needed in clinical conditions in which alloimmunization against platelets occure. Whereas antibody-typing requires serological methods, HPA-typing today is mostly done by DNA analysis. A variety of methods have been applied to HPA genotyping during the past (reveiwed in Hurd *et al.*, 2002).

Most clinical conditions with platelet alloimmunization call for HPA-typing of the patient and occasionally of some platelet donors in order to enable HPA-matched platelet transfusion. In this situation HPA-typing is needed in a flexible, low-throughput formate and PCR-SSP is currently the most widely used technique. The 5 relevant HPA systems are analyzed by applying 10 primer mixes in separate PCRs under the same conditions and evaluated after simple agarose gel electrophoresis (Figure 1).

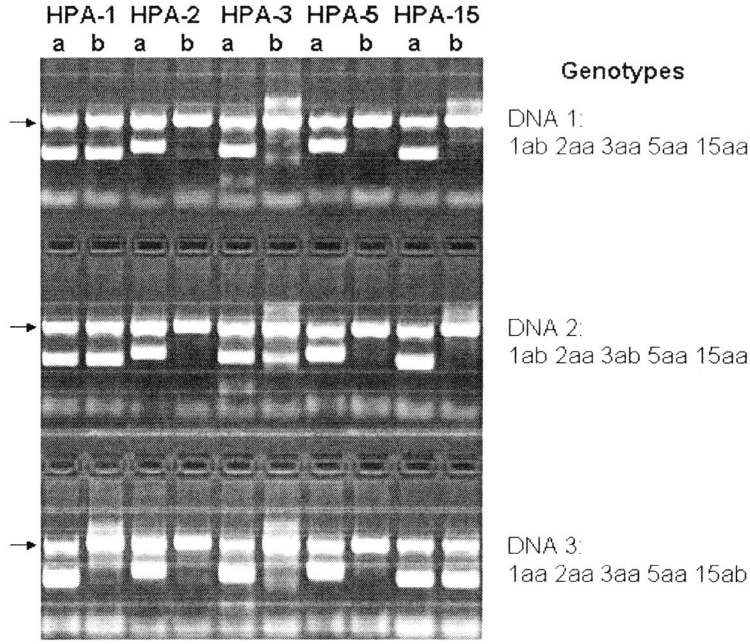

Figure 1. Genotyping of biallelic HPA systems by using PCR-SSP. Each HPA system is analyzed in two PCRs, one for the a-allele and one for the b-allele. A control fragment (black arrow) is co-amplified in each reaction and represents a conserved genomic segment independent from HPA loci.

3.2. Prenatal HPA Genotyping

NAIT differs from the Rh hemolytic disease of the newborn by occuring already in the first pregnancy. Women may become immunized to fetal alloantigens, resulting in thrombocytopenia in the infant. There is no routine screening performed to identify women at risk and newborns are often not diagnosed for NAIT until after birth with obvious signs of thrombocytopenia, such as petechiae, ecchymoses, and other types of bleeding. Therefore, the primary goal must be an early diagnosis of NAIT, preferentially in the prenatal period when alloimmunization of the mother is known. The currently preferred method of prenatal diagnosis is cord blood sampling; platelet counts and DNA for HPA-genotyping can be obtained from these samples. However, cord blood sampling is an invasive procedure, and risky especially for thrombocytopenic fetuses. Other current methods of obtaining fetal DNA, including amniocentesis, are also invasive tests and pose risks to the unborn child. The use of a PCR-SSP approach with optimized sensitivity can minimize the required volume of aminotic fluid (Bugert *et al.*, 2003b). The fetal genotype of the HPA systems could be obtained from less than 0.5 ml amniotic fluid in the early second trimester of pregnancy (Figure 2A).

The discovery of fetal DNA in maternal plasma and serum has raised the hope for noninvasive prenatal diagnostics. Recently, a PCR-SSP approach with conventional detection of PCR products in agarose gels led to the sensitive detection of a fetal point mutation in the cystic fibrosis (CF) gene in maternal plasma (Nasis *et al.*, 2004). Also the optimized PCR-SSP for HPA-genotyping is suitable to detect fetal alleles in maternal plasma in cases with reasonable amounts of fetal DNA (Figure 2B) (Bugert *et al.,* 2003b). We were able to establish a fluorescent PCR-SSP technique to further increase the sensitivity for the detection of fetal HPA-alleles in maternal plasma. After allele-specific amplification the products are detected by using a laser-assisted DNA sequencer (Figure 3). In future, the techniques might be established as routine tests to genotype SNPs or mutations using fetal DNA obtained from the maternal bloodstream instead of subjecting women to cordocentesis or amniocentesis.

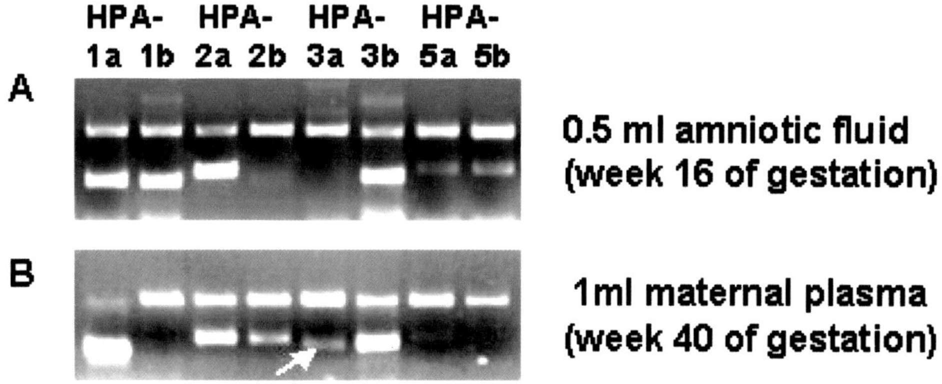

Figure 2. PCR-SSP with optimized sensitivity is suitable for obtaining (A) fetal HPA genotypes from less than 0.5 ml amniotic fluid taken in week 16 of gestation. (B) In cases with reasonable fetal DNA content in the maternal plasma, a mixed maternal/fetal HPA genotype can be detected. The fetal HPA-3a specific PCR product is marked by a white arrow (from Bugert *et al.,* 2003b).

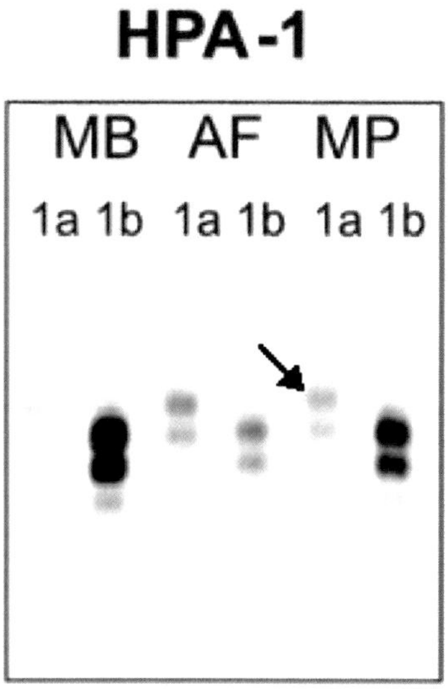

Figure 3. Detection of the fetal HPA-1a allele in maternal plasma of a NAIT case with HPA-1a alloimmunization using a fluorescent PCR-SSP. The HPA-1 genotype obtained from maternal whole blood (MB) was 1bb, the fetal genotype, HPA-1ab, was found in the amniotic fluid (AF) sample, and DNA isolated from maternal plasma (MP) revealed the specific PCR product also in the 1b-reaction (black arrow).

3.3. High Throughput Genetic Testing

In clinical laboratories where individual diagnosis of patients is the main task, high throughput genotyping techniques are not required. These approaches are rather used in pharmacogenomic or epigenetic research. In terms of routine diagnosis the high throughput techniques are currently developed for genotyping of various antigen systems, such as HLA, HPA and blood groups. Recently, several research groups described approaches for the multiplex analysis of blood groups and HPA systems either on a glass slide platform (Beiboer *et al.,* 2005; Bugert *et al.,* 2005) or other formates (Denomme *et al.,* 2005; Hashmi *et al.,* 2005). E.g., HPA loci are amplified in a single multiplex PCR with fluorescently labeled primers and hybridized to allele-specific oligonucleotide probes immobilized on glass slides. Genotypes are determined on the basis of the signal intensity ratios of a-specific vs. b-specific spots on the array (Figure 4). The technique was validated using 71 blinded DNA samples containing at least 15 aa, ab and bb genotypes for the HPA-1, -2, -3, -5 and -15 systems (Bugert *et al.,* 2005). The results of microarray-based genotyping were concordant with historical HPA genotypes obtained by PCR-SSP for all samples. The microarray-based techniques are suitable for automation with automated allelic call, which is an important criterium for a high-throughput technique. Future work may lead to a microarray platform for multiplexed high-throughput genotyping of blood groups and other antigen systems in blood banks and blood services. This would help to establish a cohort of fully typed blood donors and the supply with fully matched blood in future.

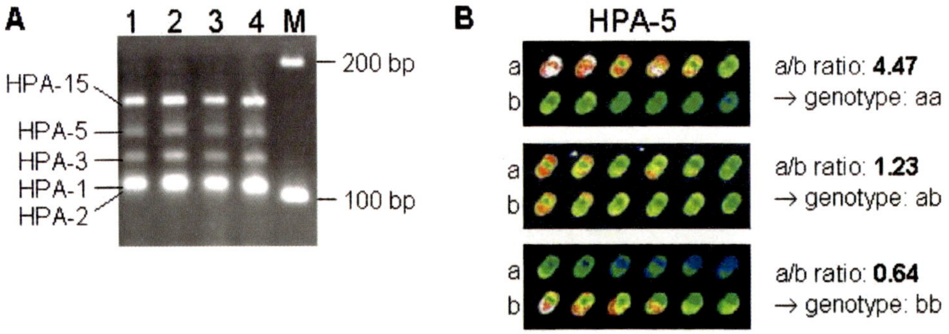

Figure 4. Microarray based genotyping of the 5 HPA loci. (A) First step is the multiplex PCR amplification of the loci. The amplicons are smaller than 200 bp. One of the primer for each locus is labelled at the 5'-end with a fluorescent dye (Cy-3). PCR products are then hybridized to arrays of allele-specific oligonucleotide probes on glass slides. (B) The HPA-5 locus is given as an example for the determination of genotypes from the ratio of hybridization signal intensity of a-allele vs. b-allele specific spots (a/b ratio).

3.4. Genotyping for further Platelet Receptor Polymorphisms

Platelet receptor polymorphisms not belonging to the HPA systems are currently not considered for routine diagnostic application. However, the gene variants are of interest for research purposes especially in epigenetic cohort studies. Therefore, fast and reliable genotyping assays are needed to analyze the polymorphisms. The analysis of SNPs in P-selectin, PSGL-1, PAR-1 or PECAM-1 is mostly done using PCR-SSP or PCR-SSO techniques (Park *et al.*, 2000; Sasaoka *et al.*, 2001; Bugert *et al.*, 2004). PCR-SSP typing of the 5 SNPs in P-selectin and the 2 SNPs in PSGL-1 can be performed under the same PCR conditions (Figure 5).

The analysis of length polymorphisms, such as the VNTRs in GPIb and PSGL-1, are simply done by PCR amplification with the use of flanking primers (Lopez *et al.*, 1992, Afshar-Kharghan *et al.*, 2001, Bugert *et al.*, 2003a). The repeat units of the PSGL-1 VNTR are 30 bp in size, and, the alleles A (16 repeats), B (15 repeats) and C (14 repeats) can be distinguished after separation on agarose gels (Figure 6).

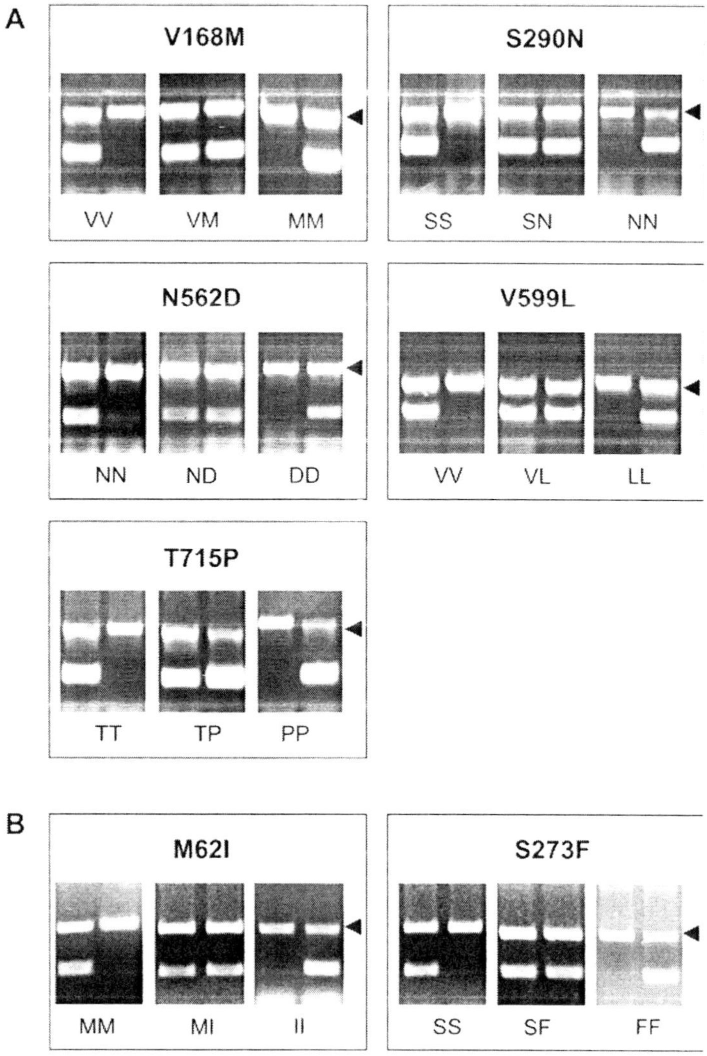

Figure 5. Results from PCR-SSP typing for the 5 SNPs in *P-selectin* (A) and the 2 SNPs in *PSGL-1*(B). Each SNP was analyzed by two allele-specific reactions each containing an allele-specific forward primer and a common reverse primer. Another primer pair directed against the *β-globin* gene served as internal control PCR product of 536 bp in size (marked by arrowhead). The genotype of a SNP was evaluated according to the presence or absence of an allele-specific PCR product and is given by the amino acid (1-letter code) encoded by each allele (from Bugert *et al.*, 2004).

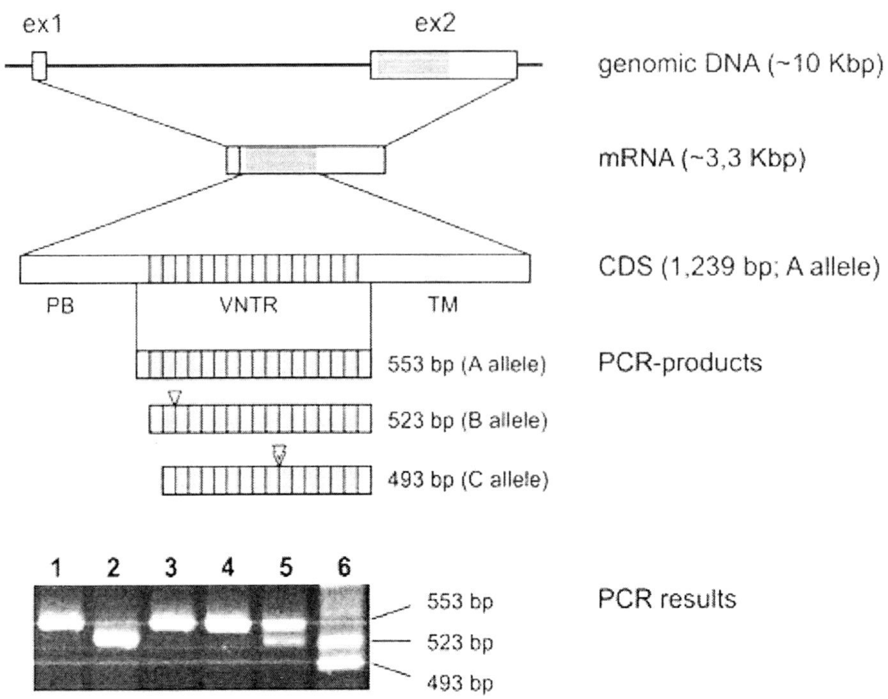

Figure 6. Genomic structure and features of the PSGL-1 gene. The entire coding region (dark gray box) for the PSGL-1 protein is located in exon 2. The position of the P-selectin binding (PB) and the transmembrane (TM) domains as well as the VNTR region in the coding sequence (CDS) is shown in the central part of the figure. The VNTR region encompassed a maximum of 16 repeat units each of 30 bp in size. Amplification of the VNTR region using flanking primers resulted in PCR products of given sizes for the A allele (16 repeats), the B allele (15 repeats; deletion of repeat unit 2) and the C allele (14 repeats; deletion of repeat units 9 and 10). Representative PCR genotyping results are given in the lower part. Lanes 1, 3 and 4: samples with A/A genotype, lane 2: B/B, lane 5: A/B, lane 6: B/C.

Conclusion

The clinical relevance of platelet receptor polymorphisms is highlighted by the HPA systems in platelet alloimmunization and the GP mutations in platelet deficiency syndromes. Other polymorphisms have been associated with various clinical phenotypes in genetic cohort studies, however to the best of our knowledge there is currently no evidence for a clear genotype-phenotype correlation of any gene variant. We have to consider that the vascular or thromboembolic diseases investigated by these studies represent complex pathomechanisms. Thus, the analysis of only one gene or one polymorphism may not be suitable to detect a strong association with the clinical phenotype. Complex investigations considering large numbers of candidate genes and their variants may be thus be needed. With regard to the role of platelets in vascular and thromboembolic disease the GPs and adhesion receptors belong to the most important candidates. The microarray-based techniques for multiplex SNP typing will be further developed to include also the non-HPA SNPs and will represent an adequate method to obtain a platelet SNP profile from patients and controls. The complex SNP typing may help to answer the questions, whether the platelet gene variants play a role in disease development and whether the platelet SNP profile can be used as a disease marker to assess the individual genetic predisposition for disease development.

References

Afshar-Kharghan V, Diz-Kücükkaya R, Ludwig EH, Marian AJ, Lopez JA. Human polymorphism of P-selectin glycoprotein ligand 1 attributed to variable numbers of tandem decameric repeats in the mucinlike region. *Blood* 2001;97:3306-3307.

Afshar-Kharghan V, Matijevic-Aleksic N, Ahn C, Boerwinkle E, Wu KK, Lopez JA. The variable number of tandem repeat polymorphism of platelet glycoprotein Ibα and risk of coronary heart disease. *Blood* 2004;103:963-965.

Apple FS, Wu AH, Mair J, Ravkil de J, Panteghini M, Tate J, Pagani F, Chrsitenson RH, Mockel M, Danne O, Jaffe AS; Committee on Standarization of Markers of Cardiac Damage of the IFCC. Future biomarkers for detection of ischemia and risk stratification in acute coronary syndrome. *Clin Chem* 2005;51:810-824.

Arnaud E, Nicaud V, Poirier O, Rendu F, Alhenc-Gelas M, Fiessinger J-N, Emmerich J, Aiach M. Protective effect of a thrombin receptor (protease-activated receptor 1) gene polymorphism toward venous thromboembolism. *Arterioscler Thromb Vasc Biol* 2000a;20:585-592.

Arnaud E, Poirier O, MONICA Members Group, Aiach M, Cambien F. The –506I/D polymorphism of the thrombin receptor (PAR-1) gene is not related to myocardial infarction in the ECTIM study. *Thromb Haemost* 2000b;84:722-723.

Balduini CL, Noris P, Giorgiani G, Martinetti M, Klersy C, Spedini P, Belletti S, MacCario R, Gusberti L, Locatelli F. Incompatibility for CD31 and human platelet antigens and acute graft-versus-host disease after bone marrow transplantation. *Br J Haematol* 1999;106:723-729.

Balduini CL, Frassoni F, Noris P, Klersy C, Iannone AM, Bacigalupo A, Giorgiani G, Di Pumpo M, Locatelli F. Donor-recipient incompatibility at CD31-codon 563 is a major risk factor for acute graft-versus-host disease after allogeneic bone marrow transplantation from a human leukocyte antigen-matched donor. *Br J Haematol* 2001;114:951-953.

Barnhart B, Ford GS, Bhushan A, Song S, Covey LR. A polymorphic CD40 ligand (CD154) molecule mediates CD40-dependent signalling but interferes with the ability of soluble CD40 to functionally block CD154:CD40 interactions. *Immunology* 2000;99:54-61.

Beiboer SH, Wieringa-Jelsma T, Maaskant-Van Wijk PA, van der Schoot CE, van Zwieten R, Roos D, den Dunnen JT, de Haas M. Rapid genotyping of blood group antigens by multiplex polymerase chain reaction and DNA microarray hybridization. *Transfusion* 2005;45:667-679.

Blankenberg S, Barbaux S, Tiret L. Adhesion molecules and atherosclerosis. *Atherosclerosis* 2003;170:191-203.

Bugert P, Hoffmann MM, Winkelmann BR, Vosberg M, Jahn J, Entelmann M, Katus HA, März W, Mansmann U, Böhm BO, Görg S, Klüter H. The variable number of tandem repeat polymorphism in the P-selectin glycoprotein ligand-1 (PSGL-1) gene is not associated with coronary heart disease. *J Mol Med* 2003a;81:495-501.

Bugert P, Lese A, Meckies J, Zieger W, Eichler H, Klüter H. Optimized sensitivity of allele-specific PCR for prenatal typing of human platelet alloantigen single nucleotide polymorphisms. *Biotechniques* 2003b;35:170-174.

Bugert P, Vosberg M, Entelmann M, Jahn J, Katus HA, Klüter H. Polymorphisms in the P-selectin (CD62P) and P-selectin glycoprotein ligand-1 (PSGL-1) genes and coronary heart disease. *Clin Chem Lab Med* 2004;42:997-1004.

Bugert P, McBride S, Smith G, Dugrillon A, Klüter H, Ouwehand WH, Metcalfe P. Microarray based genotyping for blood groups; comparison of gene array and nuclease assay techniques using HPA as a model. *Transfusion* 2005;45:654-659.

Bussel JB, Kunicki TJ, Michelson AD. Platelets: new understanding of platelet glycoproteins and their role in disease. *Hematology* 2000:222-240.

Citores MJ, Rua-Figueroa I, Rodriguez-Gallego C, Durantez A, Garcia-Laorden MI, Rodriguez-Lozano C, Rodriguez-Perez JC, Vargas JA, Perez-Aciego P. The dinucleotide repeat polymorphism in the 3'UTR of the CD154 gene has a functional role on protein expression and is associated with systemic lupus erythematosus. *Ann Rheum Dis* 2004;63:310-317.

Croft SA, Samani NJ, Teare MD, Hampton KK, Steeds RP, Channer KS, Daly ME. Novel platelet membrane glycoprotein VI dimorphism is a risk factor for myocardial infarction. *Circulation* 2001;104:1459-1463.

Denomme GA, van Oene M. High-throughput multiplex single-nucleotide polymorphism analysis for red cell and platelet antigen genotypes. *Transfusion* 2005;45:660-666.

Ertel K, Al-Tawil M, Santosos S, Kroll H. Relevance of the HPA-15 (Gov) polymorphism on CD109 in alloimmune thrombocytopenic syndromes. *Transfusion* 2005;45:366-373.

Evans RD, Perkins VC, Henry A, Stephens PE, Robinson MK, Watt FM. A tumor-associated ß1 integrin mutation that abrogates epithelial differentiation control. *J Cell Biol* 2003;160:589-596.

Ezioni A, Ochs HD. The hyper IgM sydrome – an evolving story. Pediatr Res 2004;56:519-525.

Fleiss JL, Tytun A, Ury HK. A simple approximation for calculating sample sizes for comparing independent groups. *Biometrics* 1980;36:343-346.

Fuleihan RL. Hyper IgM syndrome: the other side of the coin. *Curr Opin Pediatr* 2001;13:528-532.

Garcia-Malo MD, Corral J, Gonzalez M, Solano C, Gonzalez-Conejero R, Caballero MD, Perez R, Moraleda JM, Vicente V. Human platelet antigen systems in allogeneic peripheral blood progenitor cell transplantation: effect of human platelet antigen mismatch on platelet engraftment and graft-versus-host disease. *Transfusion* 2004;44:771 776.

Gardemann A, Knapp A, Katz N, Tillmanns H, Haberbosch W. No evidence for the CD31 C/G gene polymorphism as an independent risk factor of coronary heart disease. *Thromb Haemost* 2000;83:629.

Hashmi G, Shariff T, Seul M, Vissavajjhala P, Hue-Roye K, Charles-Pierre D, Lomas-Francis C, Chaudhuri A, Reid ME. A flexible array format for large-scale, rapid blood group DNA typing. *Transfusion* 2005;45:680-688.

Herrmann SM, Ricard S, Nicaud V, Mallet C, Evans A, Ruidavets JB, Arveiler D, Luc G, Cambien F. The P-selectin gene is highly polymorphic: reduced frequency of the Pro715 allele carriers in patients with myocardial infarction. *Hum Mol Genet* 1998;7:1277-1284.

Hurd CM, Cavanagh G, Schuh A, Ouwehand WH, Metcalfe P. Genotyping for platelet-specific antigens: techniques for the detection of single nucleotide polymorphisms. *Vox Sang* 2002;83:1-12.

Joutsi-Korhonen L, Smethurst PA, Rankin A, Gray E, Ijsseldijk M, Onley CM, Watkins NA, Williamson LM, Goodall AH, de Groot PG, Farndale RW, Ouwehand WH. The low-frequency allele of the platelet collagen signaling receptor glycoprotein VI is associated with reduced functional responses and expression. *Blood* 2003;101:4372-4379.

Juji T, Wanatabe Y, Ishikawa Y, Fujiwara K, Tonami H, Tanaka H, Satake M, Akaza T, Tadokoro K, Kodera Y, Sasazuki T, Morishima Y, Takaku F. Human platelet alloantigen (HPA)-5a/b mismatch decreases disease-free survival in unrelated bone marrow transplantation. *Tissue Antigens* 1999;54:229-234.

Kappelmayer J, Nagy B Jr, Miszti-Blasius K, Hevessy Z, Setiadi H. The emerging value of P-selectin as a disease marker. *Clin Chem Lab Med* 2004;42:475-486.

Kekomäki S, Kyllönen L, Salmela K, Koskimies S, Kekomäki R. Platelet-specific alloantigens in cadaveric renal transplantation. A prospective study. Effect of HPA-5b mismatch in acute vascular rejection of renal allografts. *Tissue Antigens* 2001;57:154-157.

Kurz M, Greinix H, Hocker P, Kalhs P, Knobl P, Mayr WR, Pober M, Panzer S. Specificities of anti-platelet antibodies in multitransfused patients with haemato-oncological disorders. *Br J Haematol* 199695:564-569.

Lee LG, Connell CR, Bloch W. Allelic discrimination by nick-translation PCR with fluorogenic probes. *Nucl Acid Res* 1993;21:3761-3766.

Leitner GC, Stiegler G, Kahls P, Greinix HT, Rabitsch W, Sillaber C, Hoecker P, Panzer S. The influence of human platelet antigen match on the success of allogenei peripheral blood progenitor cell transplantation following a reduced-intensity conditioning regimen. *Transfusion* 2005;45:195-201.

Lopez JA, Berndt MC. The GPIb-IX-V complex. In: Michelson AD, ed. *Platelets*. Academic Press; 2002:85-104.

Lopez JA, Ludwig EH, McCarthy BJ. Polymorphism of human glycoprotein Ib alpha results from a variable number of tandem repeats of a 13-amino acid sequence in the mucin-like macroglycopeptide region. Structure/function implications. *J Biol Chem* 1992;267:10055-10061.

Lopez JA, Andrews RK, Afshar-Kharghan V, Berndt MC. Bernard-Soulier syndrome. *Blood* 1998;91:4397-4418.

Lozano ML, Gonzales-Conejero R, Corral J, Rivera J, Iniesta JA, Martinez C, Vicente V. Polymorphisms of P-selectin glycoprotein ligand-1 are associated with neutrophil-platelet adhesion and with ischaemic cerebrovasculer disease. *Brit J Haematol* 2001;115:969-976.

Matsubara Y, Murata M, Sugita K, Ikeda Y. Identification of a novel point mutation in platelet glycoprotein Ibalpha, Gly to Ser at residue 233, in a Japanese family with platelet-type von Willebrand disease. *J Thromb Haemost* 2003;1:2198-2205.

Meisel C, Lopez JA, Stangl K. Role of platelet glycoprotein polymorphisms in cardiovascular diseases. *Naunyn-Schmiedberg's Arch Pharmacol* 2004;369:38-54.

Metcalfe P. Platelet antigens and antibody detection. *Vox Sang* 2004;87(suppl1):S82-S86.

Metcalfe P, Watkins NA, Ouwehand WH, Kaplan C, Newman P, Kekomaki R, de Haas M, Aster R, Shibata Y, Smith J, Kiefel V, Santoso S. Nomenclature of Human Platelet Antigens (HPA). *Vox Sang* 2003;85:240-45.

Miller JL, Cunningham D, Lyle VA, Finch CN. Mutation in the gene encoding the α chain of platelet glycoprotein Ib in platelet-type von Willebrand disease. *Proc Natl Acad Sci USA* 1991;88:4761-4765.

Nasis O, Thompson S, Hong T, Sherwood M, Radcliffe S, Jackson L, Otevrel T. Improvement in sensitivity of allele-specific PCR facilitates reliable noninvasive prenatal detection of cystic fibrosis. *Clin Chem* 2005;50:694-701.

Newman PJ. The biology of PECAM-1. *J Clin Invest* 1997;100(Suppl):S25-S29.

Novotny VM. Prevention and management of platelet transfusion refractoriness. *Vox Sang* 1999;76:1-13.

Nurden AT. Nurden P. Inherited disorders of platelet function. In: Michelson AD. ed. *Platelets*. Academic Press; 2002:681-700.

Park H-Y, Nabika T, Jang Y, Kim D, Kim H-S, Masuda J. Identification of new single-nucleotide polymorphisms in the thrombin receptor gene and their effects on coronary artery disease in Koreans. *Clin Exp Pharmacol Physiol* 2000;27:690-693.

Roldan V, Gonzalez-Conejero R, Marin F, Pineda J, Vicente V, Corral J. Short alleles of P-selectin glycoprotein ligand-1 protect against premature myocardial infarction. *Am Heart J* 2004;148:602-605.

Rozman P, Karas M, Kosir A, Labar B, Madrigal A, Middleton D, Navarrete C, Oudshorn M, Schennach H, Vitek A, Bohinjec M; Collaborative Study. Are human platelet alloantigens (HPA) minor transplantation antigens in clinical bone marrow transplantation? *Bone Marrow Transplant* 2003;31:497-506.

Russell SD, Roth GJ. Pseudo-von Willebrand disease: a mutation in the platelet glycoprotein Ibα gene associated with a hyperreactive surface receptor. *Blood* 1993;81:1787-1791.

Samara WM, Grubel PA. The role of platelet receptors and adhesion molecules in coronary artery disease. *Coronary Artery Dis* 2003;14:65-79.

Santoso S, Sachs UJ, Kroll H, Linder M, Ruf A, Preissner KT, Chavakis T. The junctional adhesion molecule 3 (JAM-3) on human platelets is a counterreceptor for the leukocyte integrin Mac-1. *J Exp Med* 2002;196:679-691.

Sasaoka T, Kimura A, Hohta S, Fukuda N, Kurosawa T, Izumi T. Polymorphisms in the platelet-endothelial cell adhesion molecule-1 (PECAM-1) gene, Asn563Ser and Gly670Arg, associated with myocardial infarction in the Japanese. *Ann NY Acad Sci* 2001;947:259-269.

Schuh AC, Watkins NA, Nguyen Q, Harmer NJ, Lin M, Prosper JYA, Campbell K, Sutherland DR, Metcalfe P, Horsfall W, Ouwehand WH. A tyrosine703serine polymorphism of CD109 defines the Gov platelet alloantigens. *Blood* 2002;99:1692-1698.

Tregouet DA, Barbaux S, Escolano S, Tahri N, Golmard JL, Tiret L, Cambien F. Specific haplotypes of the P-selectin gene are associated with myocardial infarction. *Hum Mol Genet* 2002;11:2015-2023.

Vestweber D, Blanks JE. Mechanisms that regulate the function of the selectins and their ligands. *Physiol Rev* 1999;79:181-213.

Yang J, Furie BC, Furie B. The biology of P-selectin glycoprotein ligand-1: its role as a selectin counterreceptor in leukocyte-endothelial and leukocyte-platelet interaction. *Thromb Haemost* 1999;81:1-7.

Yee DL, Bray PF. Clinical and functional consequences of platelet membrane glycoprotein polymorphisms. *Sem Thromb Hemost* 2004;30:591-600.

Zotz RB, Winkelmann BR, Nauck M, Giers G, Maruhn-Debowski B, März W, Scharf RE. Polymorphism of platelet membrane glycoprotein IIIa: human platelet antigen 1b (HPA-1b/PlA2) is an inherited risk factor for premature myocardial infarction in coronary artery disease. *Thromb Haemost* 1998;79:731-735.

Zotz RB, Klein M, Dauben HP, Moser C, Gams E, Scharf RE. Prospective analysis after coronary-artery bypass grafting: platelet GP IIIa polymorphism (HPA-1b/PIA2) is a risk factor for bypass occlusion, myocardial infarction, and death. *Thromb Haemost* 2000;83:404-407.

In: Focus on Genetic Screening Research
Editor: Sandra R. Pupecki, pp. 95-119

ISBN 1-60021-229-8

Chapter V

New Approach for Important Factor Selection from High Dimensional Bioinformatics Data

Hiro Takahashi and Hiroyuki Honda*
Department of Biotechnology, School of Engineering,
Nagoya University, Furo-cho, Chikusa-ku, Nagoya 464-8603, Japan

Abstract

Rapid advances in DNA microarray analysis technology provide new tools for establishing prognostic predictors of various diseases. In such tools, it is desired to find significant genes selectively for constructing the prognostic model and also is necessary to eliminate nonspecific genes or genes with error for construction of the model.

There are two kinds of methods, filter and wrapper approaches, for gene selection. In our previous study, we developed the projective adaptive resonance theory (PART) as filtering approach and boosted fuzzy classifier with SWEEP operator (BFCS) as a wrapper approach.

We applied PART to gene expression data in order to eliminate nonspecific genes and genes with error. Next, the genes selected by PART were subjected to our fuzzy neural network (FNN) combined with SWEEP operator method (FNN-SWEEP) for construction of a cancer class prediction models. The results in predictor model were evaluated through comparison with a conventional screening method (signal-to-noise (S2N) method) or nearest shrunken centroids (NSC) method. The FNN-SWEEP predictor with PART screening could discriminate classes of acute leukemia in blinded data with 97.1% accuracy and classes of lung cancer with 90.0% accuracy, while the predictor with S2N was only 85.3% and 70.0% or the predictor with NSC was 88.2% and 90.0%, respectively.

Next, BFCS was applied to microarray data of brain tumor (central nervous system tumor). The results in the predictor model were evaluated through comparison with a conventional screening method, such as support vector machine (SVM), FNN-SWEEP,

* Corresponding author: E-mail : honda@nubio.nagoya-u.ac.jp; phone : +81-52-789-3215; fax : +81-52-789-3214

k-nearest neighbor (kNN), multiple regression analysis (MRA), and weighted voting (WV). The accuracy of the BFCS method reached 71.3% in the 4-input models. On the other hand, SVM, FNN-SWEEP, kNN, MRA, and WV showed an accuracy of 54.3% to 66.3%, which was lower than that of BFCS.

The results have proven that PART and BFCS were superior to other conventional methods, respectively.

1. Introduction

Advances in DNA microarray technologies have made it possible to evaluate the expression levels of thousands of genes simultaneously, thereby offering a rich source of information that is potentially useful in the diagnosis and prognosis of diseases [1, 2]. There are two main methods of expression data analyses: unsupervised learning methods and supervised learning methods. The unsupervised learning methods, e.g. k-means [3], hierarchical clustering [4], self-organized maps (SOMs) [5], and Fuzzy ART [6], are designed to identify previously unrecognized classes of the diseases based on the expression pattern and the biological significance of such disease subclasses, such as prognosis, is assessed. In contrast, the supervised learning methods use training sets to specify the genes that should be clustered together [7]. However, to conduct either unsupervised or supervised analysis, it is necessary to select genes that have a strong correlation with the target phenotype, such as disease diagnosis or prognosis. This is because the performance of classification analysis can decline if a large number of genes as predictor variables are incorporated in the model.

Gene selection has been performed to screen candidate genes for modeling. There are two types of approaches—the wrapper approach and filter approach. In the former approach, genes are selected as a part of the mining algorithms, such as recursive feature elimination procedure for support vector machines (SVM) [8], fuzzy neural network (FNN) combined with SWEEP operator (FNN-SWEEP) method [1], and boosted fuzzy classifier with SWEEP operator (BFCS) method [9]. On the other hand, in the latter approach, features are selected by filtering methods, such as signal-to-noise statistic (S2N) [10], Mann-Whitney's U-test, t-test, nearest shrunken centroids (NSC) [11], and projective adaptive resonance theory (PART) [12], prior to the application of the mining algorithms.

Previously, we developed the projective adaptive resonance theory (PART) as filtering approach and boosted fuzzy classifier with SWEEP operator (BFCS) as a wrapper approach. PART has the ability to pre-select genes with small variance within a class and the ability of BFCS to build a powerful predictor by assembling multiple weak predictors.

FNN is one of the advanced artificial neural network (ANN) models, and its most attractive feature is that causality between input and output variables can be described extremely accurately as explicit IF-THEN rules obtained from the constructed model [13]. But, it takes a very long time to analyze thousands of gene expression data by FNN. In our previous work, for the purpose of dealing with thousands of genes, we developed fuzzy neural network (FNN) combined with SWEEP operator method (FNN-SWEEP method) [1]. FNN-SWEEP method has been used for microarray analysis and has proved to be precise, simple tools for predicting patients' survival [1]. However, the expression data include a huge number of genes, including experimental error, as well as nonspecific genes. In the case that artificial nonspecific gene expression data with random noise were added to real gene expression data, FNN-SWEEP method sometimes selected such artificial nonspecific genes, and the FNN model constructed by using such selected genes showed high prediction accuracy for estimation data (data not shown). This result shows that FNN-SWEEP method is sensitive to nonspecific genes and genes with error. Therefore, it is necessary to identify selectively significant genes and to eliminate nonspecific genes and genes with error before modeling. Many researchers have tried to extract significant genes from microarray data without *a priori* knowledge, for example through clustering. However, successful results have not yet been obtained by the method to eliminate genes, proposed until now. In the previous study, we applied the projective adaptive resonance theory (PART) [14] to gene expression profiles in order to eliminate nonspecific genes or genes with error [12].

In our previous study [6], Fuzzy ART based on the ART was constructed by us and applied to clustering of time course data of gene expression profiles. Cluster construction obtained by Fuzzy ART showed high reproducibility and the highest clustering robustness was obtained even when adding random noise corresponding to 2-fold change. This means that ART can select genes with a similar expression pattern with high robustness. For this reason, we applied Fuzzy ART to time course data of gene expression profiles for constructing genetic networks in our previous study [15]. Although Fuzzy ART is very useful for time course data analysis of several dimensions, this method cannot be applied for high dimensional data. PART was derived from ART to extract specific dimensions for correctness clustering of high dimensional data. Therefore, all ARTs but PART cannot extract dimensions. Cao et al. [14] reported a comparison of PART with Fuzzy ART. In their experiment, PART and Fuzzy ART were applied to artificial data sets, which have 20-dimension or 100-dimension. For both data sets, the patterns clustered correctly in the case of PART, while the patterns were very disorderly in the case of Fuzzy ART and too many clusters were generated by more rigid parameters.

In the present study, genes selected by PART were subjected to the FNN-SWEEP modeling method to construct a cancer class prediction model (predictor). The results in modeling were evaluated through comparison with those models that did not apply screening, with the conventional screening method, S2N or with NSC.

The performance of classification analysis can decline if a large number of genes as predictor variables are input in the model, because FNN-SWEEP that is based on ANN cannot avoid "curse of dimensionality". And then, we developed BFCS based on fuzzy theory and AdaBoost [16] that is, a kind of boosting algorithm. AdaBoost is the most basic boosting algorithm, and it has been used to avoid the "curse of dimensionality." BFCS also offers a simple, accurate, and high-speed system similar to AdaBoost. In addition, BFCS has a unique advantage because it can calculate the reliability index (RI) for each sample, which is not available in any other existing boosting methods.

2. Materials and Methods

2.1. Data Processing

For evaluation of PART performance, we used two kinds of gene expression profiles. The first one is the gene expression profile reported by Golub et al. [10]. These gene expression data consist of 72 bone marrow samples (47 acute lymphoblastic leukemia (ALL), and 25 acute myeloid leukemia (AML)), which were obtained from acute leukemia patients at the first time of diagnosis. RNAs prepared from bone marrow mononuclear cells were hybridized to high-density oligonucleotide microarrays (Affymetrix) containing probes for 7,129 human genes. From this data set, we selected 5,401 genes with wide expression variation, which are more than 50 of standard deviation (SD). Selection was carried to eliminate genes that had little difference in gene expression among all patients. Furthermore, these data were separated into two groups; 38 samples (27 ALL, 11 AML) and 34 samples (20 ALL, 14 AML). The former group was used as a modeling data set for constructing the class prediction model (predictor) and the latter one was used as a blinded data set for evaluating the constructed predictor.

The second one is a gene expression profile reported by Bhattacharjee et al. [17]. The gene expression data consist of 203 lung tumor samples (139 adenocarcinoma, 21 squamous cell lung carcinomas, 20 pulmonary carcinoids, 6 small cell lung carcinomas (SCLC) and 17 normal lung samples). RNAs were hybridized to high-density oligonucleotide microarrays (Affymetrix) containing probes for 12,600 human genes. From these gene expression data, we selected 3,312 genes by the same criterion mentioned above. These samples were separated into a modeling data set consisting of 198 samples and a blinded data set consisting of 5 samples (one sample from each class). FNN model was constructed from 198 modeling data and 5 blinded data were predicted. This procedure was repeated 5 more times under the condition that the same sample was not selected as blinded data. Since SCLC sample was only 6, more than 6 blinded data sets cannot be prepared. The accuracy of blinded data was calculated as the average of 6 times prediction.

For evaluation of BFCS performance, we used gene expression data set of medulloblastoma, which is a type of central nervous system (CNS) tumor reported by Pomeroy et al. [18]. Patients with medulloblastoma are treated by combinations of surgery, radiotherapy and chemotherapy. In the present data set, the following three drugs are mainly used for chemotherapy: vincristine, cisplatin and cytoxan. Therefore, by using gene selection and prognosis modeling proposed in the present study, the gene related to the treatment response can be extracted. The data set comprised 7,129 human genes (probe sets) and 60 patients from whom tumor specimens were obtained by surgery. Among these 60 patients, a few patients (16) had a short follow-up period. Therefore, we used the data of the remaining 44 patients for the construction of a 4-year survival prediction model. Of these 44 patients, 26 patients remained alive after 4 years and 18 patients had died. In this experiment, the data set was randomly partitioned into three data sets consisting of two groups: 30 or 29 patients (18 or 17 survivors, 12 dead) as a modeling data set for constructing the class prediction model (predictor) and 14 or 15 patients (8 or 9 survivors, 6 dead) as a blinded data set for evaluating the constructed predictor. We excluded those genes for which all the 44 patients showed the intensity of less than 1,000 signals [19] prior to applying the various filtering methods. Thus, 2,713 genes were selected for this study.

2.2. Evaluation of Extracted Genes by PART

To evaluate genes extracted by PART, we constructed four kinds of FNN class predictors. At first, we applied PART to the modeling data set and extracted 253 genes from the 5,401 genes of the acute leukemia data. Then, class predictor genes were selected by FNN-SWEEP from genes extracted by PART (predictor 1). Secondly, the predictor genes were selected directly from 5,401 genes without screening (predictor 2). Thirdly, the genes were selected from the 5,401 genes by S2N instead of using PART, and class predictor genes were selected from these genes (predictors 3 and 4). Fourthly, the genes were selected by nearest shrunken centroids (NSC) method, instead of using PART, and class predictor genes were selected from these genes (predictors 5). These four kinds of predictors were compared with one another with respect to their prediction accuracy of blinded data. Similarly, in the case of lung cancer data, predictor 6 was constructed by FNN-SWEEP from 387 genes (in the case of one set among six data sets) selected by PART, predictor 7 from 3,312 genes without screening, predictor 8 from the same number of genes with PART selected by MaxS2N and predictor 9 from the genes by NSC. The outline of the predictor construction is shown in Fig. 1.

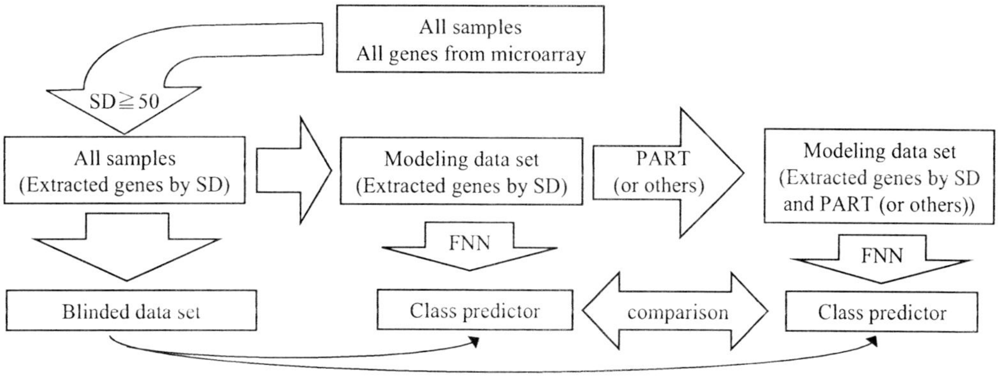

Figure 1. Outline of this study.

2.3. Projective Adaptive Resonance Theory (PART) Model

PART was proposed to find projected clusters for data sets in high dimensional spaces. The architecture is based on the well-known adaptive resonance theory (ART) developed by Carpenter et al. [20], and a major modification is provided in order to deal with the inherent sparsity in the full space of the data points (Fig. 2).

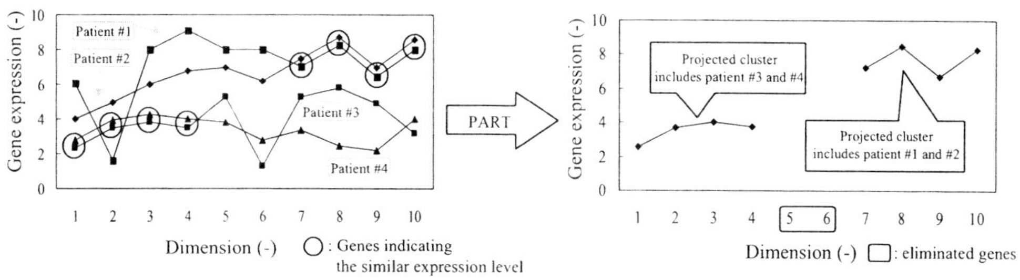

Figure 2. Function of PART clustering.

In our study, projective adaptive resonance theory (PART) [14] was used as a screening method, which enables the elimination of nonspecific dimension for clustering from high dimensional data, while conventional clustering methods cannot eliminate correctly. Briefly, the learning procedure of PART is described below.

PART includes comparison layers F_1 and clustering layers F_2, which are connected by the bottom-up weight z_{ij} and top-down weight z_{ji} (i=1, …., m: dimension number of input, j=1, …., n: category number). PART is controlled by a vigilance parameter ρ and a distance parameter σ [14]. Input vector I has dimensions corresponding to sampling genes. First of all, an input I_i itself is provided as a weight z_{ji}, and z_{ij} is set according to the following equations.

$$z_{ij} = L/(L-1+m) \tag{1}$$

where m denotes dimension number of input, L is the constant parameter, which is higher than one.

As a first step, a winner category for each input vector I is determined as follows. The function T_j of category j is defined as Eq. 2, which indicates the similarity between input vector I and weight vector z_j.

$$T_j = \sum_{i=1}^{m} z_{ij} h_{ij}(I_i, z_{ji}) \tag{2}$$

where

$$h_{ij}(I_i, z_{ji}) = h_\sigma(I_i, z_{ji}) l(z_{ij}) \tag{3}$$

where

$$h_\sigma(I_i, z_{ji}) = \begin{cases} 1 & if \ |I_i - z_{ji}| \le \sigma \\ 0 & if \ |I_i - z_{ji}| > \sigma \end{cases} \tag{4}$$

$$l(z_{ij}) = \begin{cases} 1 & if \ z_{ij} > \theta \\ 0 & if \ z_{ij} \le \theta \end{cases} \tag{5}$$

The category j that has the maximal T_j is defined as the "winner" category for input vector I.

As a next step, the category selected above is judged to follow "resonance" procedure or "mismatch reset" procedure by the function r_j defined as the following equation.

$$r_j = \sum_{i=1}^{m} h_{ij} \tag{6}$$

"Resonance" procedure is carried out if the function r_j of the winner category for input vector I is bigger than ρ; that is expressed as

$$r_j \geq \rho \tag{7}$$

The function r_j indicates the size of dimensions of projected subspace and vigilance parameter indicates its threshold. When "resonance" procedure should be done, learning of the weight vector of winner category is performed. Learning of the bottom-up weight z_{ij} and top-down weight z_{ji} is updated according to the following equations.

$$z_{ij}^{new} = \begin{cases} L/(L-1+|X|) & if\ h_{ij} = 1 \\ 0 & if\ h_{ij} = 0 \end{cases} \tag{8}$$

$$z_{ji}^{new} = (1-\alpha)z_{ji}^{old} + \alpha I_i \tag{9}$$

where α, that is the learning rate within the range from 0 to 1, was set to 0.1 which was the same value as reported by Cao et al. [14], because the change of this value was not affected for selected genes in the previous study. $|X|$ denotes the number of elements in the set $X=\{i, h_{ij}=1\}$.

Otherwise, if the function r_j of the winner category for input vector I is lower than ρ, "resonance" procedure is not done and "mismatch reset" procedure is carried out. A new category that has the next maximal T_j is chosen by Eq. 2 again. When any category cannot satisfy Eq. 7, a new category is generated according to the following equations.

$$z_{ij}^{new} = L/(L-1+|m|) \tag{10}$$

$$z_{ji}^{new} = I_i \tag{11}$$

These steps mentioned above are continued until every input vector I is assigned to any category.

We modified the above algorithm as described below. All patterns and all categories have a teacher signal. When a new category is generated, its teacher signal is decided according to the input pattern's teacher signal at that time. The "resonance" procedure is done only for the pattern having the same teacher signal as category.

Furthermore, we defined correctness ratio of clustering as the following equation.

$$correctness\ ratio = \frac{A-E}{A} \tag{12}$$

where A is all pattern number and E is mismatch pattern number with respect to cluster signal and pattern signal.

2.4. Signal-To-Noise Statistic (S2N)

The signal-to-noise statistic has been proposed to calculate the weight of genes for weighed voting algorithm as a binary class predictor by Golub et al. [10]. This statistic is defined as the following equation.

$$S2N = \left| \frac{\mu_{class1} - \mu_{class2}}{\sigma_{class1} + \sigma_{class2}} \right| \tag{13}$$

where μ is the average of log gene expression for each class, and σ is the standard deviation of log gene expression for each class.

We calculate the S2N value of all class pairs for each gene in a multi-class prediction, and then the biggest S2N value of all pairs is defined as MaxS2N value for the gene.

$$MaxS2N = \max \left\{ \left| \frac{\mu_{class_i} - \mu_{class_j}}{\sigma_{class_i} + \sigma_{class_j}} \right| : i \in C, j \in C, i \neq j \right\} \tag{14}$$

where C is the set of classes.

2.5. Nearest Shrunken Centroids (NSC)

NSC method has been proposed to identify minimal subsets of the genes by Tibshirani et al. [11]. In this method, the mean expression of each gene within each class is calculated. Then it shrinks these centroids toward the overall mean for that gene by a fixed quantity, threshold Δ. The NSC value is defined as the following equation.

$$NSC = sign(d_{ik})\,(|\,d_{ik}\,| - \Delta)_{+} \tag{15}$$

where i is each gene, k is each class, + is positive part (if $t > 0$ then $t_+ = t$ else $t_+ = 0$), and d_{ik} is defined as the following equation.

$$d_{ik} = \frac{\bar{x}_{ik} - \bar{x}_i}{m_k \cdot (s_i + s_0)} \tag{16}$$

where $\bar{x}_{ik}$ is centroids of ith gene in the class k, $\bar{x}_i$ is overall centroids of ith gene, s_i is the pooled within-class standard deviation for gene i, s_0 is a positive constant value (we set it equal to the median value of the s_i over the set of the genes), and m_k is defined as the following equation.

$$m_k = \sqrt{\frac{1}{n_k} + \frac{1}{n}} \tag{17}$$

where n_k is sample number in the class k and n is all sample number.

In this study, the FNN-SWEEP method is then applied to the genes that survive the thresholding. The threshold Δ was optimized by cross-validation in modeling data set.

2.6. Fuzzy Neural Network (FNN)

Fuzzy modeling is a method for describing the characteristics of a system using fuzzy inference rules [21]. The distinguishing feature of the method is that it can express complex nonlinear systems linguistically by IF-THEN rules or fuzzy IF-THEN rules. It is extremely difficult to identify the rules and tune the shape of the fuzzy membership function. Horikawa et al. [22] proposed the use of fuzzy neural networks (FNNs) for resolving this problem. They are based on the use of the back-propagation algorithm [23], and they can automatically acquire the rules and the tuned fuzzy membership function. However, the learning of the networks is time consuming and the network cannot avoid the "curse of dimensionality."

2.7. Fuzzy Neural Network (FNN) Combined with SWEEP Operator Method (FNN-SWEEP)

The fuzzy neural network (FNN) combined with SWEEP operator method (FNN-SWEEP method) was also applied for model construction. The FNN-SWEEP method was originally proposed by Noguchi et al. [24] and was modified by Ando et al. [1] to manage microarray data. FNN has three types of weight parameters (w_c, w_g, and w_f) [22] as shown in Fig. 3. If w_c and w_g are fixed, FNN can be treated as multiple linear regression model in which w_f is variable parameters. Therefore, w_f was easily optimized without training. In the FNN-SWEEP method, only parameter w_f was optimized by the SWEEP operator method during the feature selection step. After the input combinations were determined, FNN models with the selected input combinations were optimized using a backpropagation algorithm on model construction step. In the backpropagation algorithm, the number of epochs was set to 5,000, and the learning rate was set to 0.1; these values are the same as those reported by Ando et al. [1].

2.8. Boosted Fuzzy Classifier with SWEEP Operator (BFCS)

Boosting was proposed by Schapire [25], and thus far, several derivative boosting algorithms [16, 26, 27] have been developed. Boosting is useful for class prediction using high dimensional inputs and is very fast algorithms. However, it is difficult to evaluate the reliability of prediction results since the conventional boosting uses the C4.5 algorithm [28] as a weak learner.

In the previous study, we developed a boosted fuzzy classifier with SWEEP operator (BFCS) method [9] on the basis of AdaBoost [16], which is the most basic boosting algorithm. This method enables the evaluation of reliability of the predictions for each patient. On the other hand, it is difficult to evaluate the reliability of the predicted results of the conventional boosting.

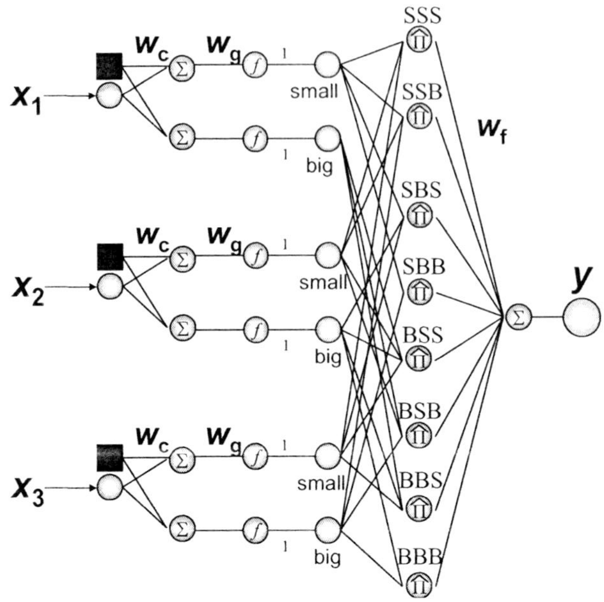

Figure 3. 3-input type-I FNN model

Figure 4 shows the structure of BFCS. BFCS is composed of 1-input type I fuzzy neural network (FNN) models [22]. In the present study, 1-input FNN models were used as weak learners in the BFCS model, and they were combined by connection weights, which were determined by the AdaBoost algorithm. FNN has three types of weight parameters (w_c, w_g, and w_f) [22]. In this study, parameter w_g is a constant value (= 2.0 × ln ((1.0 + 0.995)/(1.0 - 0.995))) [22], and w_c is a threshold that has the best odds ratio in the case that only one input was used. w_c and w_g were determined; w_f was calculated by the SWEEP operator method [1]. Briefly, the construction procedure of BFCS was as follows.

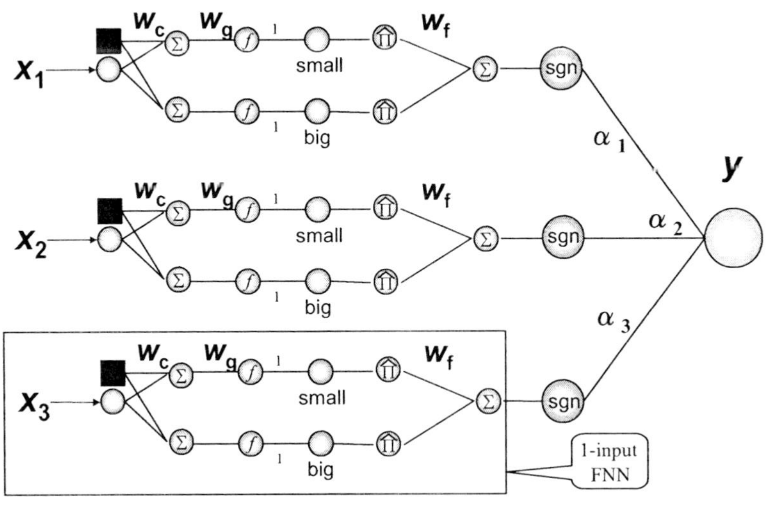

Figure 4. Concept of the BFCS model.

(A) Algorithm of BFCS

Step 1: Construction of weak learners

1-input FNN models were constructed for each input.

Step 2: Selection of weak learners and calculation of connection weights for construction of BFCS models

(i) Initialization

Weight w_i for each sample (i = 1,...., p = sample number) is set to $1/p$.

Weight f_j for each model (j = 1,...., m: model number) is set to 0.

(ii) Calculation of weighted error rates and model selection

$$\varepsilon_j = \sum_{i:h_j(x_i)\neq y_i}^{p} w_i \tag{18}$$

$$\varepsilon_s = \underset{f_j=0}{\arg\min}(\varepsilon_j) \tag{19}$$

where $h_j(x_i)$ indicates output of jth weak learner model when i^{th} sample was inputted into the model and y_i indicates the teacher signal of ith sample. ε_s, which is the smallest among ε_j, which has $f_j = 0$, is selected for model selection.

(iii) Update of weights

$$\alpha_t = \frac{1}{2}\ln\frac{1-\varepsilon_s}{\varepsilon_s} \tag{20}$$

$$w_i^{new} = w_i^{old} \bullet \exp(-\alpha \bullet y_i \bullet h_s(x_i))/Z \tag{21}$$

$$f_j = \begin{cases} 1 & \text{if } j\text{th model has more than one inputs which were the same inputs in the } s\text{th model} \\ 0 & \text{if } j\text{th model has no inputs which were the same inputs in the } s\text{th model} \end{cases} \tag{22}$$

where α_t indicates the connection weight of the tth selected model and Z is the normalized factor defined such that $\sum_{i=1}^{p} w_i^{new}$ equals 1.

(iv) Repeat (ii) to (iii) until the following stopping condition is satisfied or all weights f_j become equal to 1.

Finally, constructed model is defined by the following equation.

$$F(x) = \sum_{t=1}^{N_{attribute}} \alpha_t \times h_t(x) \tag{23}$$

where $F(x)$ indicates constructed model. $h_t(x)$ indicates t^{th} selected weak learner.

The genes with the 1st to the 10th highest accuracies were used as the first inputs for the construction of the 10 combination models by boosting.

(B) Reliability index for BFCS

Reliability index (RI) based on fuzzy inference has been proposed to evaluate the result of class prediction by Huang and Li [29]. We have developed a reliability index for BFCS (RI_{BFCS}) by modifying RI for boosting.

We modified RI equations as follows:

$$RI_{BFCS} = \begin{cases} \text{INTEGER}(\text{diff}_{BFCS} \cdot 10) + 1 & 0 \le \text{diff}_{BFCS} < 0.9 \\ 10 & \text{diff}_{BFCS} \ge 0.9 \end{cases} \quad (24)$$

where

$$\text{diff}_{BFCS} = \frac{\sum_{t}^{T} \left\{ \alpha_t \bullet \underset{v \in M_t}{\arg\min}(\text{diff}_v) \right\}}{\sum_{t}^{T} \alpha_t} \quad (25)$$

where T indicates the number of weak learners in the BFCS model, M_t indicates set of input variables in t^{th} selected weak learner, indicates the connection weight of the tth selected model in the construction of BFCS models, and $diff_v$ is defined by the following equation.

$$\text{diff}_v = u_{highest}(x_v) - u_{next\ highest}(x_v) \quad (26)$$

where v indicates the vth input in the BFCS model and $u(x_v)$ indicates the grade of the fuzzy membership function when the vth input x_v was inputted. It is defined by the following equation [22].

$$u_v = \frac{1}{1 + \exp\{-w_g (x_v + w_c)\}} \quad (27)$$

RI_{BFCS} is calculated for each example. Here, the greater RI_{BFCS} the sample has, the more reliable its prediction.

2.9. k-Nearest Neighbor (kNN)

The k-nearest neighbor (kNN) methods are based on a *distance function* for pairs of tumor samples, such as the Euclidean distance. The kNN proceeds as follows to classify blind data set observations on the basis of the modeling data set. For each patient in the blind data set (a) finding the k closest patients in the modeling data set, and (b) predicting the class by majority vote; that is, choosing the class that is most common among those k neighbors. The number of neighbors k=3 was used because a similar cross-validation accuracy of model was obtained in the modeling data set for various k.

2.10. Multiple Regression Analysis (MRA)

The multiple regression analysis (MRA) is one of conventional methods. The MRA is a concerned with describing and evaluating the relationship between a patient's outcome and gene expression. MRA models are used to help us predict a patient's outcome by using gene expression data.

2.11. Weighted Voting (WV)

The Weighted voting (WV) method was originally proposed by Golub et al. [10] to manage microarray data. The weights of each gene were calculated by the signal-to-noise. The linear models of one gene were assembled with gene weight.

2.12. Support Vector Machine (SVM)

The support vector machine (SVM) was originally proposed by Vapnik and Chervonenkis [30] and is used to avoid the "curse of dimensionality." SVM is superior to many other conventional methods and is frequently used in bioinformatics [31-33]. In this study, the SVM-LIGHT software package [34] was used. It was modified, and a PIM function was added to select a combination of inputs. In the present study, the regulatory parameter c was the default value of SVM_LIGHT ((avg.(input vector)^2)$^{-1}$). A linear kernel was used because a similar cross-validation accuracy of model was obtained in the modeling data set for various kernels.

2.13. Model Construction with Parameter Selection

The parameter increasing method (PIM) [24] was used to select input combinations for model construction of FNN-SWEEP, SVM, kNN, MRA, and WV. This was done as follows:

First, we predicted the subtype of each sample by using the prediction model with a single input. Prediction models for each probe were constructed in a series, and all the probes were ordered based on the accuracy of the constructed models. In the next step, the probe having the highest accuracy level was used for constructing a combination model.

Second, we selected a partner probe for the probe selected in the first step in order to increase the prediction accuracy. To accomplish this, we constructed a 2-input model in which a ranked probe was designated as input 1, and input 2 (partner probe) was selected to provide the highest training accuracy while applying FNN-SWEEP (or SVM, kNN, MRA, and WV) and PIM to the modeling data. By repeating this step, a combination of $N_{\text{attribute}}$ candidate probes was identified for use as input probes in the model construction.

Finally, combinations of $N_{attribute}$ probes, i.e., from the first to the $N_{attribute}$th probe were evaluated. We constructed $N_{attribute}$ predictor models, beginning with one input using only the first-selected probe to $N_{attribute}$ inputs using all the $N_{attribute}$ probes. The predictor models were specifically constructed by using a backpropagation algorithm for FNN-SWEEP or quadratic programming for SVM. The performance of the prediction models was evaluated by applying them to the blinded data set.

The genes with the 1st to the 10th highest accuracies were used as the first inputs for the construction of the 10 combination models by PIM.

3. Results and Discussion

3.1. Case of PART for Acute Leukemia and Lung Cancer

3.1.1. Identification of Optimal Gene Number of PART Screening

The number of genes selected by PART is controlled by a vigilance parameter ρ and a distance parameter σ. For the vigilance parameter, we selected the smallest ρ for which the highest correctness ratio was obtained for clustering, with a fixed distance parameter. For example, when the distance parameter was 2.7 in the analysis of acute leukemia, the vigilance parameter became 1 in the range of 1 to 5,401 of integers so as to achieve the clustering correctness ratio of 100.0%, and then the number of selected genes was 253. In order to identify the optimal gene number in PART screening, the distance parameter should also be surveyed. For this purpose, an FNN-SWEEP method was carried out for gene selection and model construction, and ten independent FNN class predictors were constructed. As shown in our previous study [1], FNN modeling was carried out by a parameter increasing method. Therefore, the number of input units was optimized during this procedure. In the case of acute leukemia, ten FNN models with one input were constructed. In the case of lung cancer, ten FNN models with four inputs were constructed. We calculated the average of the model accuracy.

The results are shown in Table 1 and Fig. 5 for acute leukemia, and Fig. 6 for lung cancer. Excess elimination decreased the accuracy of the model, which may be caused by the elimination of important genes. In the case of acute leukemia, the model constructed by selecting 244 genes showed 92.1% of modeling data accuracy, selecting 253 genes increased its accuracy to 93.1%. In the case of lung cancer, the model constructed by selecting 352 genes showed 92.4% of modeling data accuracy, and selecting 387 genes increased its accuracy to 94.7%. Based on these results, we selected 253 genes and 387 genes as the lowest gene number, respectively.

Table 1. Comparison of PART parameters and optimal gene numbers

For acute leukemia						
Distance parameter (-)	2.5	2.6	2.7	2.8	2.9	100.0
Vigilance parameter (-)	1	1	1	1	1	1
Selected gene number (-)	233	244	253	262	271	5401
For lung cancer*						
Distance parameter (-)	0.9	1.0	1.1	1.2	1.3	100.0
Vigilance parameter (-)	30	26	47	49	47	1
Selected gene number (-)	320	352	387	419	449	3312

* the parameters for one set of six lung cancer data sets.

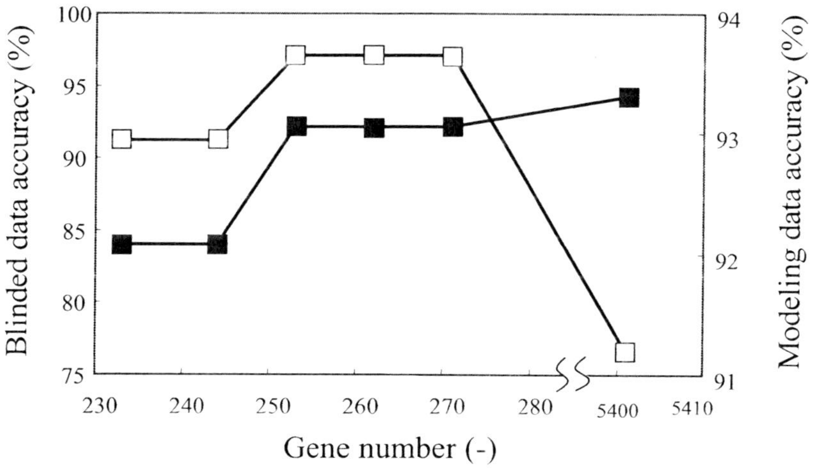

Figure 5. Accuracy on various gene numbers for acute leukemia. Solid line with open squares indicates accuracy for blinded data. Solid line with closed squares indicates accuracy for modeling data.

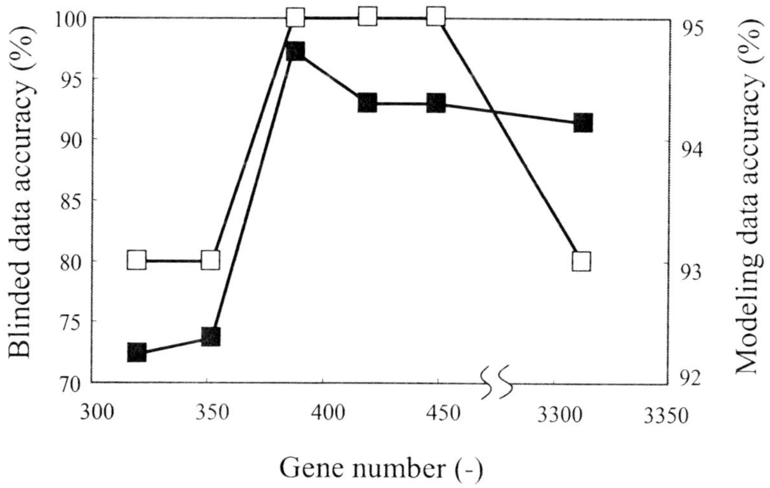

Figure 6. Accuracy on various gene numbers for one set of six lung cancer modeling data sets. Solid line with open squares indicates accuracy for blinded data. Solid line with closed squares indicates accuracy for modeling data.

3.1.2. Comparison of the Performance of FNN-SWEEP Class Predictor with PART and other Screening Method

Performance of the class predictor constructed with PART screening was investigated. For comparison, class predictors were constructed using the S2N ranking method, which has been frequently used by many researchers or NSC method. We also constructed a predictor without screening. Class predictors, which can correctly classify not only modeling data but also new data, should be constructed. Therefore, the performance of the predictors was compared for accuracy using blinded data, which was never used for modeling. The average accuracy for blinded data using ten independent FNN models was calculated. In the case of acute leukemia, the top 50 genes were selected by means of S2N, because 50 genes had been used for analysis by weighted voting method in the report of Golub et al. [10], and those were used for FNN-SWEEP analysis. In addition, the top 253 genes were also selected by S2N so as to select the same number of genes as those resulting from PART screening, and those were used for FNN-SWEEP analysis. In the case of multiple class data of lung cancer, the top 387 genes were selected by S2N to evaluate the discrimination accuracy of the predictor constructed by the FNN-SWEEP method. The top 387 genes were defined by MaxS2N. Furthermore, the case using NSC was also compared.

As shown in Table 2, the predictor with PART screening showed significantly high performance. All predictors for acute leukemia showed high modeling ability for modeling data (approximate 93% of accuracy). However, for blinded data, predictor 1 with PART screening showed 97.1% accuracy, while predictor 2 without screening was only 76.5% accurate. In addition, predictors resulting from the S2N method (predictors 3 and 4) and NSC method (predictor 5) showed only 85.3% and 88.2% of accuracy for blinded data, respectively. In the case of acute leukemia, we prepared 34 samples of blinded data. Therefore, 5 samples out of 34 were predicted incorrectly using the S2N method, while only 1 was incorrect for PART. It should be noted that the accuracy never increased with predictor 7 using 253 candidate genes from S2N. These data mean that PART screening can select genes with the similar expression pattern with high robustness.

Table 2. Accuracy of class prediction for acute leukemia and lung cancer

Model	Gene Screening	Discrimination accuracy*1	
		Modeling data set*2	Blinded data set
For acute leukemia			
Predictor 1	PART	93.1%	97.1%
Predictor 2	Nothing	93.3%	76.5%
Predictor 3	top 50 by S2N	92.9%	85.3%
Predictor 4	top 253 by S2N	93.1%	85.3%
Predictor 5	NSC	93.7%	88.2%
For lung cancer			
Predictor 6	PART	93.7±0.8%	90.0±9.3%
Predictor 7	Nothing	93.4±0.4%	80.0±0.0%
Predictor 8	MaxS2N	90.4±1.9%	70.0±9.3%
Predictor 9	NSC	94.0±0.4%	90.0±9.3%

*1 one blinded data set was prepared for acute leukemia, and 6 blinded data sets were prepared for lung cancer. For lung cancer, the average of accuracy from 6 models corresponding to 6 blinded data sets was listed.

*2 the value was obtained from 3-fold cross validation for acute leukemia and 5-fold cross validation for lung cancer.

In the case of acute leukemia, it should be noted that only 10 significant genes were used for class prediction, since FNN models with one input were constructed, while 50 genes were necessary for the same analysis in the conventional method using weighted voting method [10]. This is a superior feature from the viewpoint of a biological experiment. If the number of genes of interest is small, the researcher could easily investigate the gene expression level by RT-PCR or histological staining of expressed protein on the specimen in order to know the relationship between gene expression and cancer classification.

For multiple class data of lung cancer, four kinds of FNN class prediction models, such as the model with PART screening, without screening, and with S2N screening were constructed. As shown in Table 2, predictor 6 with PART screening showed 93.7% accuracy for modeling data and 90.0% for blinded data and predictor 9 with NSC screening showed 94.0% accuracy for modeling data and 90.0% for blinded data. On the other hand, predictor 7 without screening was 93.4% and 80.0% accurate, respectively. Predictor 8 with MaxS2N screening was also only 90.4% accurate for modeling data and 70.0% for blinded data, respectively. The reason why blinded data accuracy of MaxS2N was low, is that the gene number selected by MaxS2N was matched to one by PART, and this shows that it is possible that PART can condense more significant genes than MaxS2N. In all cases, 10 FNN models were constructed for prediction and for predictor 6 only 17 independent significant genes were used for class prediction, since FNN models with four inputs were constructed.

3.1.3. Comparison of Genes Used in a Predictor with PART, with other Screening

The genes selected by the FNN-SWEEP predictor with PART screening were compared with those used in the predictor with NSC, S2N or without screening. The deviation of gene expression level of genes selected by these four methods was investigated. In this examination, all patients are divided into two classes, such as AML or ALL. Average expression levels of genes used in the predictor with PART screening were calculated for each patient class. The SD of gene expression was calculated against 10 genes in each class. Averages of SDs are also shown in Fig. 7. The average SD of the expression level of genes used in the predictor with PART screening in the lower class was significantly smaller than that of the higher class. The value of the SD for the predictor with PART screening was 1/2.3, 1/2.5 or 1/3.6 times less than that with NSC, S2N or without screening. Although in the case of higher class a higher SD was obtained in the predictor with PART, the difference is not so big and 1.2, 1.2 or 1.3 times higher SD were obtained compared with that with NSC, S2N or without screening. Low SD values mean that signals of those genes have less noise and the genes show similar expression level in a class. If genes with a high SD were used in a predictor, the model constructed will show low accuracy for blinded data, although high accuracy may be obtained for modeling data. Actually, as shown in Table 2, discrimination accuracy of all models for the modeling data set was approximate 93%. However, the accuracy of predictor 1 with PART screening was 97.1% for blinded data, while the accuracy of the other models was significantly lower than this. The high accuracy will be the result of low SD values in the lower class.

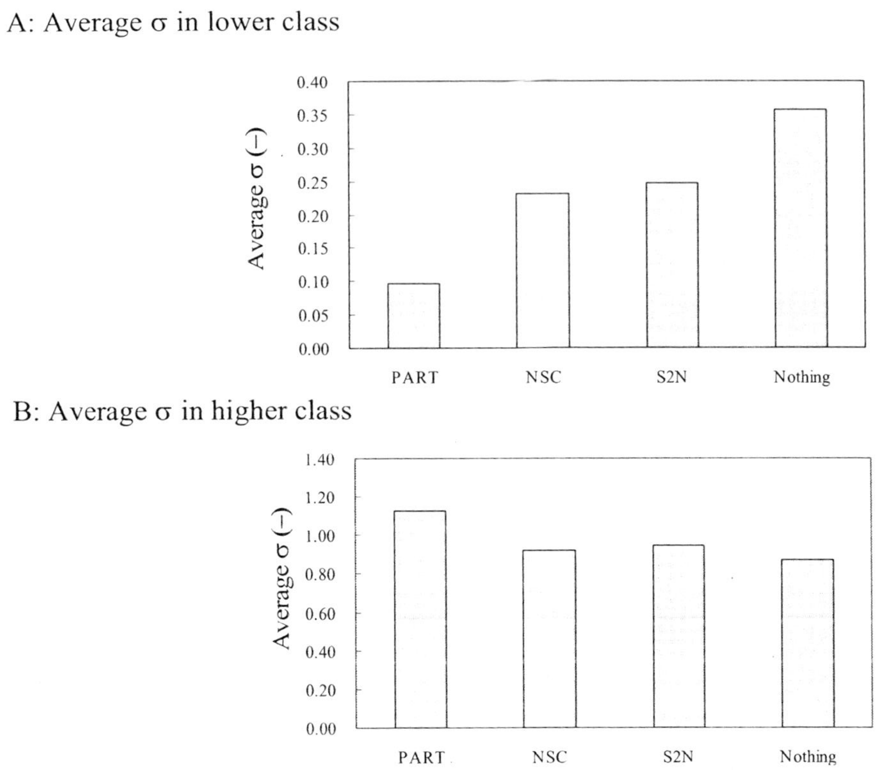

Figure 7. Comparison of average of each parameter for 10 genes of predictors indicates average of the SD. Average and SD of gene expression data of all patients were normalized to become zero and one, respectively.

Genes that were part of the FNN class predictor for acute leukemia are shown in Table 3. Ten genes in the FNN model with PART screening included five genes from the top 50 genes selected by the S2N method as reported by Golub et al. [10]. The zyxin (X95735) gene was also selected in other predictors derived from either the S2N method or without screening, and the gene was commonly used in the first FNN model in other methods. This result means that the zyxin gene was very important for discriminating acute leukemia and PART did not eliminate this important gene. Furthermore, CD36 antigen gene expression was selected only in the predictor with PART screening. CD36 antigen has been reported by Valet et al. [35] as a high risk AML marker gene. This may be due to the fact that the SD in the lower class of the predictor with PART was the smallest among the four predictors as shown in Fig. 7.

For clustering of lung cancer, we investigated the presence of previously reported marker genes among the genes selected in the constructed FNN predictors. As shown in Table 4, 17 genes were selected in the predictor with PART screening. Among those, five genes were marker genes reported by Bhattacharjee et al. [17]: tumor protein 63 kDa with strong homology to p53 and keratin 5 for squamous cell lung carcinoma (SQ), adv. glycosylation end product-sp. receptor for normal lung (NL), and ISL1 transcription factor and insulinoma-associated 1 for neuroendocrine tumor (NE). In addition to these genes, two genes among the remaining 12 genes, *chromogranin A* and *chromogranin C*, were reported to be marker genes for NE by Lamberts et al. [36]. One gene, ceruloplasmin, was also reported to be a marker gene for AD and SQ by Wang et al. [37]. These findings suggest that the genes selected in the predictor with PART screening may identify new marker genes and also suggest that the conventional method may miss the selection of important genes for class prediction.

Table 3. The genes used in FNN-SWEEP class predictor for acute leukemia

Accession No.	Description	Genes included in top 50 by S2N
D88422	CYSTATIN A	
M27783	ELA2 Elastatse 2, neutrophil	
M27891	CST3 Cystatin C (amyloid angiopathy and cerebral hemorrhage)	*
M31166	PTX3 Pentaxin-related gene	
M77142	NUCLEOLYSIN TIA-1	
M84526	DF D component of complement (adipsin)	*
M91432	ACADM Acyl-Coenzyme A dehydrogenase	*
M96326	Azurocidin	*
M98399	CD36 antigen (collagen type I receptor, thrombospondin receptor)	
X95735	Zyxin	*

Table 4. The genes used in FNN-SWEEP class predictor for one set of six lung cancer modeling data sets

Accession No.	Description	Selected by Bhattacharjee et al.	Marker genes
AB018279	KIAA0736 protein		
AB029010	KIAA1087 protein		
AF032906	Cathepsin Z		
D21267	Synaptosomal-associated protein, 25kD		
L18983	Protein tyrosine phosphatase, receptor type, N		
M13699	Ceruloplasmin		*1
M21389	Keratin 5	*	
M25756	Chromogranin C		*2
M91211	Human receptor for advanced glycosylation end products (RAGE) mRNA	*	
M93119	Insulinoma-associated 1	*	
U03749	Chromogranin A		*2
U07559	ISL1 transcription factor	*	
U48437	Human amyloid precursor-like protein 1 mRNA		
U62317	Hypothetical protein 384D8_6		
Y00757	Secretory granule, neuroendocrine protein 1		
Y16752	Secretagogin		
Y16961	Tumor protein 63 kDa with strong homology to p53	*	

*1 The marker gene reported by Wang et al. [37]

*2 The marker genes reported by Lamberts et al. [36]

3.2. Case of BFCS for Central Nervous System (CNS) Tumor

3.2.1. Comparison of the Performance of BFCS Class Predictor and other Wrapper Methods

In the previous study, BFCS was applied to cancer diagnosis using clinical data [38] for breast cancer and proteomic pattern data [39] of mass spectrometry (MS) for ovarian cancer. In this study, BFCS was applied to microarray data [18] of brain tumor (central nervous system tumor). The performance of BFCS models as class predictors was investigated. For comparison, many models of various wrapper approaches, such as SVM, FNN-SWEEP, k-nearest neighbor (kNN), multiple regression analysis (MRA), and weighted voting (WV), were constructed. The performance of the predictors was compared based on the accuracy by using a blinded data set that was not used for modeling. By using 10 independent class predictor models, the average accuracy for blinded data set was calculated for the CNS and leukemia data sets.

The results for CNS data are shown in Fig. 8. The inputs used in the predictors were 4-input model. As shown in Fig. 8, the BFCS method clearly showed high performance when compared with the other methods. The BFCS predictor could discriminate classes in blinded data with 71.3% accuracy by using 4-input models, while SVM was only 66.3%, FNN-SWEEP was 58.7%, kNN was 60.4%, MRA was 54.3%, and WV was 62.5%, respectively. This result implies that BFCS can efficiently combine 1-input FNN models as a weak learner. Furthermore, BFCS can select the genes by itself without parameter increasing method (PIM). On the other hand, some feature selection methods like PIM are necessary for other methods to select genes.

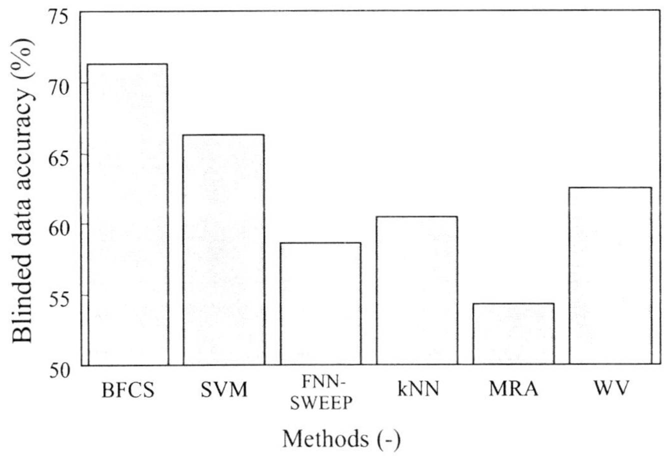

Figure 8. Comparison of accuracies on various wrapper approaches for CNS data set.

3.2.2. Comparison of Genes Used in BFCS Predictors and FNN-SWEEP Predictors for CNS Data

We compared FNN-SWEEP and BFCS to investigate numerical character of the genes selected by BFCS. Both FNN-SWEEP and BFCS are based on FNN. The one-gene predictors were constructed for each gene from 2nd input to 4th input in the two methods. And then, average modeling accuracies of one-gene predictors for ten combinations, were calculated

(Fig. 9). The BFCS genes used as one-gene predictors showed clearly higher accuracy than FNN-SWEEP ones, as shown in Fig. 9. The average modeling accuracies of the genes from 2nd to 4th were 83.3%, 77.3% and 79.0% for BFCS, and 72.0%, 68.7% and 65.7% for FNN-SWEEP, respectively. The PIM method was used in the FNN-SWEEP. This method is very useful to select input combination that shows high accuracy by combining low accuracy inputs. But the application of PIM to high dimensional data, such as microarray data, may cause overfitting. On the other hand, the boosting used in the BFCS is the method that can construct high-accuracy predictor by combining one-gene predictors. Thus one-gene predictors with low-accuracy are hardly selected. It may be for this reason that BFCS showed high performance.

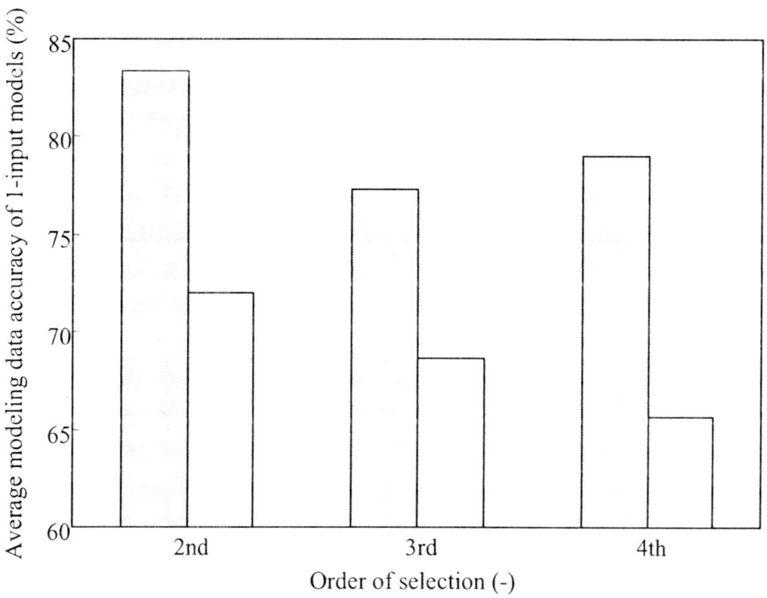

Figure 9. Average modeling accuracy of 1-input models between BFCS and FNN-SWEEP.

In this experiment, 10 combinations of 4-input models were constructed. Three genes from 2nd to 4th in 4-input were selected as combination of genes for each method. The modeling accuracies when these three genes were used alone as one-gene predictors, were calculated for 10 combinations. White bars and grey bars mean accuracies of FNN-SWEEP and BFCS, respectively.

3.2.3. Comparison of Selected Marker Genes with Known Prognostic Genes

We investigated the presence of previously reported prognostic marker genes among the genes selected in the 10 constructed combinations of 4-inputs BFCS models. There were a total of 40 genes in ten models. Some genes were selected several times. In the case of BFCS, 19 genes among 40 genes are independent, as shown in Table 5. Two genes among these 19 genes were reported to be prognostic markers for cancer: The *USP4* (*UNPH*) gene was reported by Frederick et al. [40] to be a low-risk marker gene. The *LIF* gene was reported by Park et al. [41] to be a low-risk marker gene. *LIF* induces growth arrest and differentiation of cells. These observations accurately matched with low or high gene expression of the above-mentioned three marker genes, as shown in Table 5. These findings suggest that the BFCS method may be used to identify new marker genes.

Table 5. The genes used in BFCS class predictor for one set of three CNS modeling data sets

Accession no.	Description	Marker genes
AB000460	Chromosome 4 open reading frame 8	
D30756	Neighbor of BRCA1 gene 1	
HG1112 -HT1112	Ras-Like Protein Tc4	
HG884 -HT884	Oncogene E6 - Ap, Papillomavirus	
L07648	MAX interactor 1	
L10333	Reticulon 1	
L11669	Tetracycline transporter-like protein	
L18983	Protein tyrosine phosphatase, receptor type, N	
L33243	Polycystic kidney disease 1 (autosomal dominant)	
M73547	Chromosome 5 open reading frame 18	
M86546	Pre-B-cell leukemia transcription factor 1	
S69265	ELAV (embryonic lethal, abnormal vision, Drosophila)-like 3 (Hu antigen C)	
S76475	Neurotrophic tyrosine kinase, receptor, type 3 (TrkC)	
U20657	Ubiquitin specific protease 4 (proto-oncogene)	*1
U77604	Microsomal glutathione S-transferase 2	
U78521	Aryl hydrocarbon receptor interacting protein	
X13967	Leukemia inhibitory factor (cholinergic differentiation factor)	*2
X80763	5-hydroxytryptamine (serotonin) receptor 2C	
Y09616	Carboxylesterase 2 (intestine, liver)	

*1The marker gene reported by Frederick et al. [40]
*2The marker genes reported by Park et al. [41]

4. Conclusion

We developed the projective adaptive resonance theory (PART) as filtering approach and boosted fuzzy classifier with SWEEP operator (BFCS) as a wrapper approach. PART and BFCS were applied to microarray data of some cancers. The results in the predictor model were evaluated through comparison with various conventional screening methods. These results suggest that the PART and BFCS methods have a high potential to function as a new method of important factor selection from high dimensional bioinformatics data, such as DNA microarray, mass spectrometry (MS), and two-dimensional polyacrylamide gel electrophoresis (2D-PAGE) and those are attractive tool for the diagnosis of patients.

References

[1] Ando T, Suguro M, Hanai T, Kobayashi T, Honda H, Seto M. Fuzzy neural network applied to gene expression profiling for predicting the prognosis of diffuse large B-cell lymphoma. *Jpn. J. Cancer Res.* 2002;93:1207-1212.

[2] Kebriaei P, Anastasi J, Larson RA. Acute lymphoblastic leukaemia: diagnosis and classification. *Best. Pract. Res. Clin. Haematol.* 2002;15:597-621.

[3] Somogyi R. Making sense of gene-expression data. Cambridge, England: *Elsevier Trends Journal;* 1999.

[4] Eisen MB, Spellman PT, Brown PO, Botstein D. Cluster analysis and display of genome-wide expression patterns. *Proc. Natl. Acad. Sci. USA* 1998;95:14863-14868.

[5] Tamayo P, Slonim D, Mesirov J, Zhu Q, Kitareewan S, Dmitrovsky E, et al. Interpreting patterns of gene expression with self-organizing maps: methods and application to hematopoietic differentiation. *Proc. Natl. Acad. Sci. USA* 1999;96:2907-2912.

[6] Tomida S, Hanai T, Honda H, Kobayashi T. Analysis of expression profile using fuzzy adaptive resonance theory. *Bioinformatics* 2002;18:1073-1083.

[7] Brown MPS, Grundy WN, Lin D, Cristianini N, Sugnet CW, Furey TS, et al. Knowledge-based analysis of microarray gene expression data by using support vector machines. *Proc. Natl. Acad. Sci. USA* 2000;97:262-267.

[8] Guyon I, Weston J, Barnhill S, Vapnik V. Gene selection for cancer classification using support vector machines. *Machine Learning* 2002;46:389-422.

[9] Takahashi H, Honda H. A new reliable cancer diagnosis method using boosted fuzzy classifier with SWEEP operator method. *J. Chem. Eng. Jpn.* 2005;38:763-773.

[10] Golub TR, Slonim DK, Tamayo P, Huard C, Gaasenbeek M, Mesirov JP, et al. Molecular classification of cancer: class discovery and class prediction by gene expression monitoring. *Science* 1999;286:531-537.

[11] Tibshirani R, Hastie T, Narasimhan B, Chu G. Diagnosis of multiple cancer types by shrunken centroids of gene expression. *Proc. Natl. Acad. Sci. USA* 2002;99:6567-6572.

[12] Takahashi H, Kobayashi T, Honda H. Construction of robust prognostic predictors by using projective adaptive resonance theory as a gene filtering method. *Bioinformatics* 2005;21:179-186.

[13] Noguchi H, Hanai T, Honda H, Harrison LC, Kobayashi T. Fuzzy neural network-based prediction of the motif for MHC class II binding peptides. *J. Biosci. Bioeng.* 2001;92:227-231.

[14] Cao Y, Wu J. Projective ART for clustering data sets in high dimensional spaces. *Neural Netw.* 2002;15:105-120.

[15] Takahashi H, Tomida S, Kobayashi T, Honda H. Inference of common genetic network using fuzzy adaptive resonance theory associated matrix method. *J.Biosci.Bioeng.* 2003;96:161-167.

[16] Freund Y, Schapire RE. A decision-theoretic generalization of online learning and an application to boosting. *J. Comput. System Sci.* 1997;55:119-139.

[17] Bhattacharjee A, Richards WG, Staunton J, Li C, Monti S, Vasa P, et al. Classification of human lung carcinomas by mRNA expression profiling reveals distinct adenocarcinoma subclasses. *Proc. Natl. Acad. Sci. USA* 2001;98:13790-13795.

[18] Pomeroy SL, Tamayo P, Gaasenbeek M, Sturla LM, Angelo M, McLaughlin ME, et al. Prediction of central nervous system embryonal tumour outcome based on gene expression. *Nature* 2002;415:436-442.

[19] Park DJ, Vuong PT, de Vos S, Douer D, Koeffler HP. Comparative analysis of genes regulated by PML/RAR alpha and PLZF/RAR alpha in response to retinoic acid using oligonucleotide arrays. *Blood* 2003;102:3727-3736.

[20] Carpenter GA, Grossberg S. A massively parallel architecture for a self-organizing neural pattern recognition machine. *Computer Vision, Graphics, and Image Processing* 1987;37:54-115.

[21] Takagi T, Sugeno M. Fuzzy identification of systems and its applications to modeling and control. *IEEE* T. *Syst. Man Cyb.* 1985;15:116-132.

[22] Horikawa S, Furuhashi T, Uchikawa Y. On fuzzy modeling using fuzzy neural networks with the back-propagation algorithm. *IEEE T. Neural Networ.* 1992;3:801-806.

[23] Rumelhart DE, Hinton GE, Williams RJ. Learning representations by back-propagating errors. *Nature* 1986;323:533-536.

[24] Noguchi H, Hanai T, Takahashi W, Ichii T, Tanikawa M, Masuoka S, et al. Model construction for quality of beer and brewing process using FNN.(in Japanese) *Kagaku Kogaku Ronbunshu* 1999;25:695-701.

[25] Schapire RE. The strength of weak learnability. *Machine Learning* 1990;5:197-227.

[26] Friedman J, Hastie T, Tibshirani R. Additive logistic regression: a statistical view of boosting. *Ann. Stat.* 2000;28:337-407.

[27] Freund Y. An adaptive version of the boost by majority algorithm. *Machine Learning* 2000; 43:293-318.

[28] Quinlan JR. C4.5: programs for machine learning. *San Mateo: Morgan Kaufmann Publishers;* 1993.

[29] Huang Y, Li Y. Prediction of protein subcellular locations using fuzzy k-NN method. *Bioinformatics* 2004;20:21-28.

[30] Vapnik VN, Chervonenkis A. A note on one class of perceptrons. *Automation and Remote Control* 1964;25:821-837.

[31] Cheng BY, Carbonell JG, Klein-Seetharaman J. Protein classification based on text document classification techniques. *Proteins* 2005;58:955-970.

[32] Wang Y, Tetko IV, Hall MA, Frank E, Facius A, Mayer KFX, et al. Gene selection from microarray data for cancer classification a machine learning approach. *Comput. Biol. Chem.* 2005;29:37-46.

[33] Han LY, Cai CZ, Ji ZL, Chen YZ. Prediction of functional class of novel viral proteins by a statistical learning method irrespective of sequence similarity. *Virology* 2005;331:136-43.

[34] Joachims T. Making large-scale SVM learning practical. Cambridge: *MIT Press;* 1999.

[35] Valet G, Repp R, Link H, Ehninger A, Gramatzki MM. Pretherapeutic identification of high-risk acute myeloid leukemia (AML) patients from immunophenotypic, cytogenetic, and clinical parameters. *Cytometry B Clin. Cytom.* 2003;53:4-10.

[36] Lamberts SW, Hofland LJ, Nobels FR. Neuroendocrine tumor markers. *Front. Neuroendocrinol.* 2001;22:309-339.

[37] Wang KK, Liu N, Radulovich N, Wigle DA, Johnston MR, Shepherd FA, et al. Novel candidate tumor marker genes for lung adenocarcinoma. *Oncogene* 2002;21:7598-7604.

[38] Wolberg WH, Mangasarian OL. Multisurface method of pattern separation for medical diagnosis applied to breast cytology. *Proc. Natl. Acad. Sci. USA* 1990;87:9193-9196.

[39] Petricoin EF, Ardekani AM, Hitt BA, Levine PJ, Fusaro VA, Steinberg SM, et al. Use of proteomic patterns in serum to identify ovarian cancer. *Lancet* 2002;359:572-577.

[40] Frederick A, Rolfe M, Chiu MI. The human UNP locus at 3p21.31 encodes two tissue-selective, cytoplasmic isoforms with deubiquitinating activity that have reduced expression in small cell lung carcinoma cell lines. *Oncogene* 1998;16:153-165.

[41] Park JI, Strock CJ, Ball DW, Nelkin BD. The Ras/Raf/MEK/extracellular signal-regulated kinase pathway induces autocrine-paracrine growth inhibition via the leukemia inhibitory factor/JAK/STAT pathway. *Mol. Cell Biol.* 2003;23:543-554.

In: Focus on Genetic Screening Research
Editor: Sandra R. Pupecki, pp. 121-139
ISBN 1-60021-229-8

Chapter VI

Application of Microsatellite Marker Analysis

Lan Kluwe
Department of Neurosurgery, University Hospital Eppendorf,
Martinistrasse 52, Hamburg, Germany

Abstract

Microstellites are DNA sequences consisting of short repeats of highly variable number. Hundreds thousands of such sequences have been found throughout the human genome. Microsatellites are extremely powerful as markers in distinguishing alleles at linked loci and thus is widely used for linkage studies, segregation analysis and determining allele loss or gain in tumors. In this chapter, several extended applications of microsatellites are markers will be introduced:

1. Using multiple highly informative microsatellites for screening of large (megabase) deletions.
2. Using information obtained from loss of heterozygosity analysis for
   a. tracing parental origin of alleles bearing de novo mutations
   b. determining mutation-carrying alleles for presymptomatic / prenatal diagnosis for at-risk individuals
   c. finding cellular origin of tumor cells
   d. verification of the identity of cells cultured from tumors.

Introduction

Originally, satellite DNA was a term for large centromeric tandem repeats because they were found in a "satellite" peak in density gradient fractionation of sheared DNA. Later, shorter tandem repeats (10-30 bp) were analogically called minisatellites and finally, sequences containing very short repeats of 1 to 7 bp were termed microsatellites.

Microsatellites are also known as short tandem repeats or simple sequence repeats. An example of microsatellite containing "CA" repeats is given as following:

Name: D22S430
Location: chromosome 22, bp 28936331 - 28936414
agtgcagcagatccttcttaaccatagtatttatttggaattttttctaattgcagatg
taaattgtgcagaatgcagaaaggtaggactcacctgtgcatgcgtgtgtgcctctatg**c**
acacacacacacacacacacacacacacaacgggaggaaaagccaagatcatttatagct
cactaccctgggtagccctggttaacccctcagagtacctccctccagtctttcatttat
tccctgaaggggttgattccaagtgctgtaagatagggaaacccagggaggtgatgggga
ccagggaagtcgggatggccttggggagaaggtgagattcagactcagtcttaaaagagg

In the above sequence, the number of the "CA" repeats is 15. However, this number varies to large extent in each allele and thus give rise to polymorphism in length. Heterozygosity of this microsatellite D22S430 has been estimated to be 0.68 which means that 68% of examined individuals have two different alleles. As will described below, two bands on a gel or two peaks on a electrogram are then visible for this microsatellite amplified. In contrast, if a marker is uninformative, it´s two alleles will have the same length which then appear as a single band on a gel or a single peak in the electrogram. In the case of D22S430, 8 alleles have been reported. The frequency of each alleles can be obtained by examine a number of individuals (table 1).

Table 1. Multiple alleles and their frequency of microsatellite D22S430

Allele	1	2	3	4	5	6	7	8
Frequency	0.1	0.01	0.17	0.04	0.01	0.04	0.12	0.51

A single nucleotide polymorphism (SNP) has only two alleles. For example, an SNP allele for an A/C polymorphism can only be either A or C. Such markers are often not informative, that means, an individual often has A or C on both alleles. In contrast, as shown in the above example, a microsatellite may have multiple alleles. This makes microsatellites very informative and extremely powerful as markers for distinguishing alleles at linked loci.

Microsatellites are highly abundant in human genome, more than 600,000 microsatellites have already been found. Di-nucleotide repeats, for example CA and GT repeats, are the most abundant microsatellites.

Finding Microsatellites

Microsatellites can be found, for example, by the Genome Browser Gateway of the UCSC Genome Bioinformatics service (http://genome.ucsc.edu/cgi-bin/hgTracks). Not all microsatellites are highly informative. Generally, microsatellites with more repeats are more informative. For example, a di-nucleotide repeats should have more than ten times of the repeats. Once a microsatellite has been found in the region of interest, the sequence flanking the repeats can be obtained. This sequence should be checked using BLAST (www.ncbi.nih.gov/BLAST/) to ensure unique localization of the microsatellite. Sequences with homologous elsewhere should be avoided.

Analyzing Microsatellites

A microsatellite can be analysed by PCR amplification and subsequent sizing. The primers for amplification of a microsatellite are chosen in the sequences flanking the repeats. To design the primer, for example, the program "Primer3" (http://bioinformatics.weizmann .ac.il/cgi-bin/primer/primer3.cgi) can be used. Some microsatellites are known as markers and can be found at www.ensembl.org/Homo_sapiens/ index.html. At this website and in addition links, the localization of the microsatellites in genome ensemble, primers and further information can be found. Sequences of these markers can also be obtained, enabling of designing alternative primers. This is necessary, for example, if a smaller or larger size is desired, or the given ones do not work well.

Amplified microsatellites can be analysed either manually using denaturing sequencing gels or an automated fluorescence sequencer, for example one of the Genetic Analyzer Series from Applied Biosystems. Fluorescence-labelling and use of an automated sequencer enables analysis of multiple microsatellites simultaneously in one load. Though multiplex-PCR is also possible, we found that the procedure of optimizing the condition is laborious. We thus prefer amplifying each microsatellite separately and pool the products together for subsequent sizing. Individual microsatellite can be distinguished either by different size-ranges or by different dyes for labelling (an example in Fig. 1). The sequencers from Applied Biosystems can distinguish five dyes. While one dye is reserved for the internal size standard, 4 different dyes can be chosen for labelling each microsatellite. More than 10 microsatellites can thus be easily separated in on run. Another advantage of using fluorescence sequencers is that absolute size of each microsatellite-allele can be called using internal size standard (Fig. 2). This enables comparison with data obtained in previous or later runs. A typical run on a ABI Sequencer takes 30 min. Thus, for urgent cases, a sample can be genotyped with multiple microsatellite markers within hours.

However, some laboratories may not have access to a fluorescence sequencer. An alternative is manually sizing of the amplified microsatellites. One of the two primers for a microsatellite has to be labelled with ^{33}P before the PCR amplification. The amplified markers can then be separated on a conventional denaturing sequencing gel. This is much laborious and need at least two days of turn around.

Though it is also possible to stain the amplified microsatellites in gel with Sybr or silver, both sensitivity and resolution are much poor than end-labelling. This is because that the staining methods visualize two strands which usually have different contents of GC and AT, and thus migrate differently.

Neither is non-denaturing condition recommended for analyzing microsatellite markers. An informative microsatellite has two different alleles. This leads to formation of two homoduplex and two heteroduplex, all with different migrating-velocity. Four bands will thus be visible. In addition, as will be described below, an amplified di-nucleotide microsatellite has multiple fragments with 2-bp length differences. The interpretation and scoring are thus very difficult and often impossible.

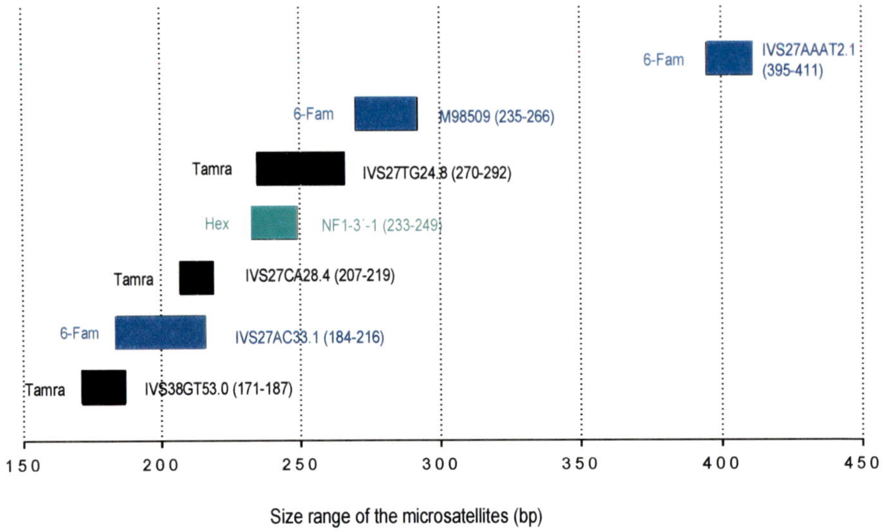

Figure 1. An example of choosing dyes for labelling primers. Notice that microsatellites in overlapping size-range are labelled with different fluorescence dyes so that they can be pooled after the amplification and analysed in one load on a sequencer.

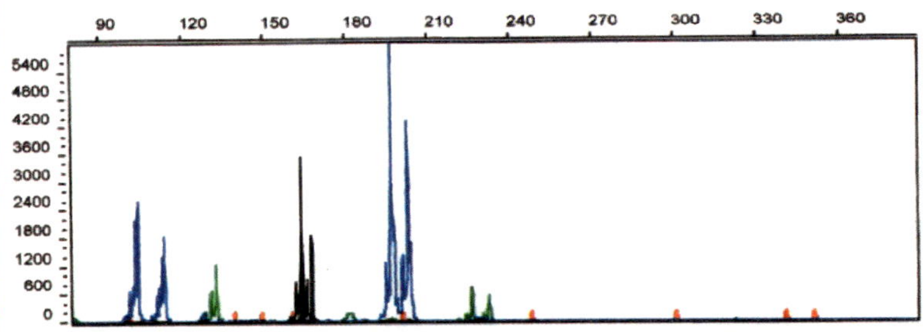

Dye/Sample Peak	Minutes	Size	Peak Height	Peak Area	Data Point
B, 2	11.69	103.21	2683	26010	3188
B, 8	11.97	112.93	1855	14573	3263
B, 18	14.16	196.21	6136	48938	3860
B, 20	14.31	202.13	4339	34595	3903
G, 5	12.50	132.18	1297	7508	3409
G, 15	14.92	225.70	753	4391	4067
G, 18	15.07	231.83	601	3686	4109
Y, 6	13.30	163.50	3566	18703	3627
Y, 8	13.40	167.41	1873	9631	3654
R, 1 •	11.60	100.00	215	1054	3163
R, 2 •	12.69	139.00	231	1246	3460
R, 3 •	12.96	149.00	235	1268	3534
R, 4 •	13.21	160.00	246	1410	3603
R, 7 •	14.26	200.00	242	1374	3888
R, 9 •	16.72	300.00	243	1496	4558
R, 10 •	17.62	340.00	242	1595	4804
R, 11 •	17.85	350.00	249	1649	4868
R, 12 •	18.99	400.00	233	1708	5179
R, 13 •	20.06	450.00	241	1826	5469
R, 14 •	20.89	490.00	228	1911	5697
R, 15 •	21.07	500.00	232	1897	5746

Figure 2. Electrogram of 5 microsatellites sized in one load. The red peaks are internal size standard run simultaneously with the amplified microsatellites. These size standard enable calculation of the absolute size of the microsatellites which are given in following table. The GeneScan program also calculate the peak area of each signal.

Interpretation of Pattern of an Amplified Microsatellite

If we label only one of the two primers for a microsatellite, only one strand will be visible for each allele. Even though, the pattern of a microsatellite, especially those of di-nucleotide microsatellite, is often complex. Correct interpretation largely depends on experience. This is because an amplified di-nucleotide microsatellite often have several fragments differing in 2-, 4- and 6-bp, giving a "ladder" pattern.

An example of di-nucleotide microsatellite sized on a fluorescence sequencer is given in Fig. 3. In Fig. 3A, the microsatellite D22S929 is homozygous, thus only one allele. Nevertheless, it is not represented by only one peak, but multiple peaks differing in 2-bp. The intensity of the peaks decrease as the size getting smaller with steps of 2-bp. The true allele for the microsatellite is the largest one on the right which also has the highest intensity. The peaks with smaller size on the left are called shadow or shoulder. Depending on microsatellite and amplification, one or more such shadow fragments are visible for a di-nucleotide microsatellite.

An explanation for the "ladder" a di-nucleotide microsatellite may be "slipping" during the amplification, in which the strand in synthesis slips one di-nucleotide repeat forward. The magnitude of the shadow effect varies from microsatellite to microsatellite and from amplification to amplification. In microsatellites consisting of tri- and tetra-nucleotide repeats, the "slipping" effect is much less prominent.

The shadow fragments often make the interpretation of pattern of an amplified microsatellite difficult, especially if two alleles differ only by a single repeat unit, in the case of di-nucleotide repeat, 2-bp (Fig. 3B). As will be described later, the interpretation of partial loss of heterozygosity in a tumor is even more challenging. Two rules are helpful for interpretation of the pattern of sized microsatellites:

(1) the right allele is the largest and
(2) the right allele has the highest intensity.

In addition, fragments of 1-bp larger is also present in some cases. This is due to the overhand of polymerase which adds a single nucleotide to the 3′-end of newly synthesized strand. Each of the real allele and shadow fragments of an amplified microsatellite have corresponding fragments with overhang. Overhang fragments can be distinguished from shadow fragments because (1) an overhang fragment differs only 1-bp from that without overhang while a shadow fragment differs 2-bp from the real allele fragment, (2) an overhang fragment is larger than that without overhang while a shadow fragment is smaller than its corresponding real allele fragment. If the overhang is very prominent leading to complex pattern of a microsatellite, it can be removed by treating the amplified microsatellite with 3′-5′exonuclease, for example Klenow polymerase.

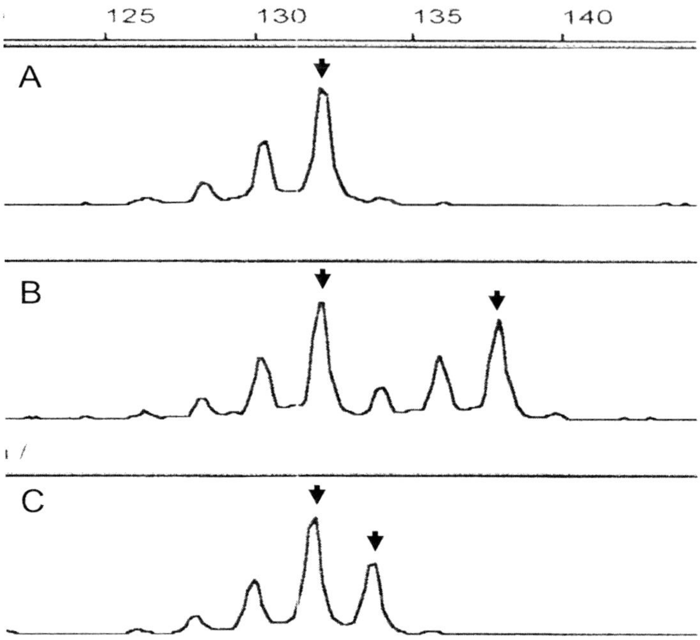

Figure 3. Patterns of a di-nucleotide microsatellite D22S929 sized on a fluorescence sequencer. **A**: one allele consisting of a main peak (array) and three smaller shadow peaks. The 3rd one, which is 6 bp smaller than the main peak, is almost invisible. **B**: two alleles differing in 6 bp. Each has three multiple peaks, making the pattern complex. The most right ones in each group represents the right alleles (arrays). Other smaller ones are shadow fragments. Notice that the signal of the smaller allele is always stronger than that of the larger allele. Two factors contribute to this phenomenon: (1) smaller fragments amplify more efficiently than larger ones and (2) the signal of one of the shadow fragments of the larger allele (in the case of B, the 3rd one) is added to the right signal of the smaller allele. **C**: two alleles differing in 2-bp. To interpret this pattern, we need to keep two rules in mind: (1) shadow signals are always less intensive than the right allele and (2) they are always shorter than the true allele and thus appear only on the left site. Therefore we know that in the pattern presented in C, the most right signal must be a true allele. But why is it less strong than the 2nd peak? This is because the 2nd peak is not a shadow peak, but the smaller allele. The first shadow signal of the larger allele is added to the right allele of the smaller one.

Extended Applications of Microsatellites

Standard applications of microsatellites are not the main focus of this chapter. Instead, several extended applications will be introduced.

1. Screening for Large Deletions

At least 5.5% of all pathogenic genetic alterations in human are large genomic deletions or duplications (Krawczak and Cooper 1997). Traditionally, Southern hybridization and FISH are methods for screening such large alterations. However, these techniques are laborious and require special facility which are not available in many laboratories. Furthermore, Southern hybridization requires high quality of DNA and FISH on interphase nuclei needs fresh specimen. This kind of samples are often not available and these conventional methods are thus not suitable for screening large number of cases. In contrast, several recently developed semi-quantitative PCR-based gene dosage assays including multiple ligation-dependent probe amplification (MLPA) and quantitative multiplex PCR of short fluorescent fragments need only small amount of DNA. Promising results using these techniques have been reported for genes such as *MSH2*, *MSH1*, *BRCA1* and *p53*. One disadvantage of these assays is the laborious optimization of primers and amplification conditions for each target in each laboratory. For example, MLPA primers are not chemically synthesized and their application depends on commercial availability of optimized primer sets.

We used multiple highly informative microsatellites as a simple, inexpensive and effective alternative strategy for screening large deletions and duplications. The principle of the screen is illustrated Fig. 4. Consider a total of 105 cases to be screened. Five of them have large deletions, while other 100 do not. We do not know which patients have the deletion but can identify most of the 100 cases with no deletions using multiple microsatellite markers. This is because if a sample is heterozygous for one microsatellite, deletion of the regions covering this marker can be excluded.

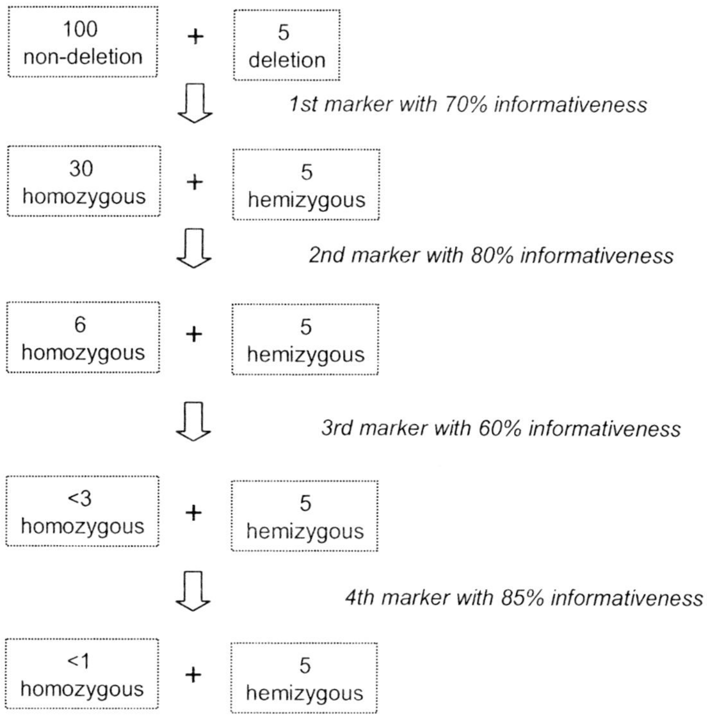

Figure 4. Illustration of the principle of screening large deletions using multiple microsatellite markers

To understand the principle in another way, lets see how it works if we analyze the microsatellites sequentially. First, the 105 cases are genotyped with a microsatellite with a 70% heterozygousity, which means that 70% individuals will have two alleles for this marker. Statistically, 70 cases will be heterozygous and thus have two alleles for this marker. These 70 cases can be excluded for a large deletion, because a case of deletion can not have two different alleles. The other 30 cases will be homozygous and the 5 deletion cases will be hemizygous for the marker. Both groups show one allele for this marker and can not be distinguished from each other. To further exclude non-deletion cases, we genotype these 35 cases with a second marker of 80% heterozygousity. Statistically, 24 (80% x 30) will be heterozygous for this 2nd marker and can thus be identified as not having large deletion, leaving 6 cases homozygous and 5 deletions cases hemizygous. Similarly, a third screen exclude 60% of the 6 cases left from the 2nd screen. After the 4th marker, less than 1 cases of non-deletion will be left. The total of 6 cases can then be further analysed by either FISH or Southern hybridization to identify the 5 deletion cases.

Table 2 gives an example of calculating the probability of an individual being homozygous for multiple markers. The probability of an individual being homozygous for multiple markers is very low, less than 0.2% for 5 markers. Therefore, if an individual is homo/hemizygous for multiple markers, the probability is high that he/she has a heterozygous deletion covering the markers. Due to linkage disequilibrium, homozygosity for multiple markers may be more frequent than we would expect from calculation. Nevertheless, our recent study demonstrated that more than 90% of non-deletion cases can be excluded based on heterozygosity for at least one marker (Kluwe et al. 2004).

Table 2. Probability for homozygosity for multiple highly informative microsatellite markers

Number of marker	Heterozygosity	Probability of an individual being heterozygosity for at least one marker	Probability of an individual being homozygosity for all markers
1	70%	70%	30%
2	80%	94%	6% (= 30% x 20%)
3	60%	98%	2.4% (= 6% x 40%)
4	85%	99%	0.36% (= 2.4% x 15%)
5	50%	99.8%	0.18% (= 0.36% x 50%)
6	75%	99.96%	0.04% (= 0.18% x 25%)
7	65%	99.99%	0.01% (= 0.04% x 35%)

In the study, we genotyped 433 neurofibromatosis patients with 7 markers. Heterozygosity for at least one marker was found in 399 (92%) cases (Fig. 5A). These cases could thus be excluded from deletions. Thirty four patients were homo- or hemizygous for all 7 markers (an example in Fig. 5B). Twenty three out of these 34 candidate could be further screened by FISH. Twenty two (96%) were confirmed to have deletions of the examined region. Only one (4%) case did not have deletion. Homozygosity for all 7 markers in this cases was thus coincident.

In our case, the analysis of microsatellites was done using a fluorescence sequencer. It was thus easy to screen 7 markers at once, because we could pooled them together and load in one run. For laboratories with no access to an fluorescence sequencer, microsatellite markers can be analyzed sequentially. This has the advantage, that the number of samples to be analysed decreases significantly after each round of analysis (table 2).

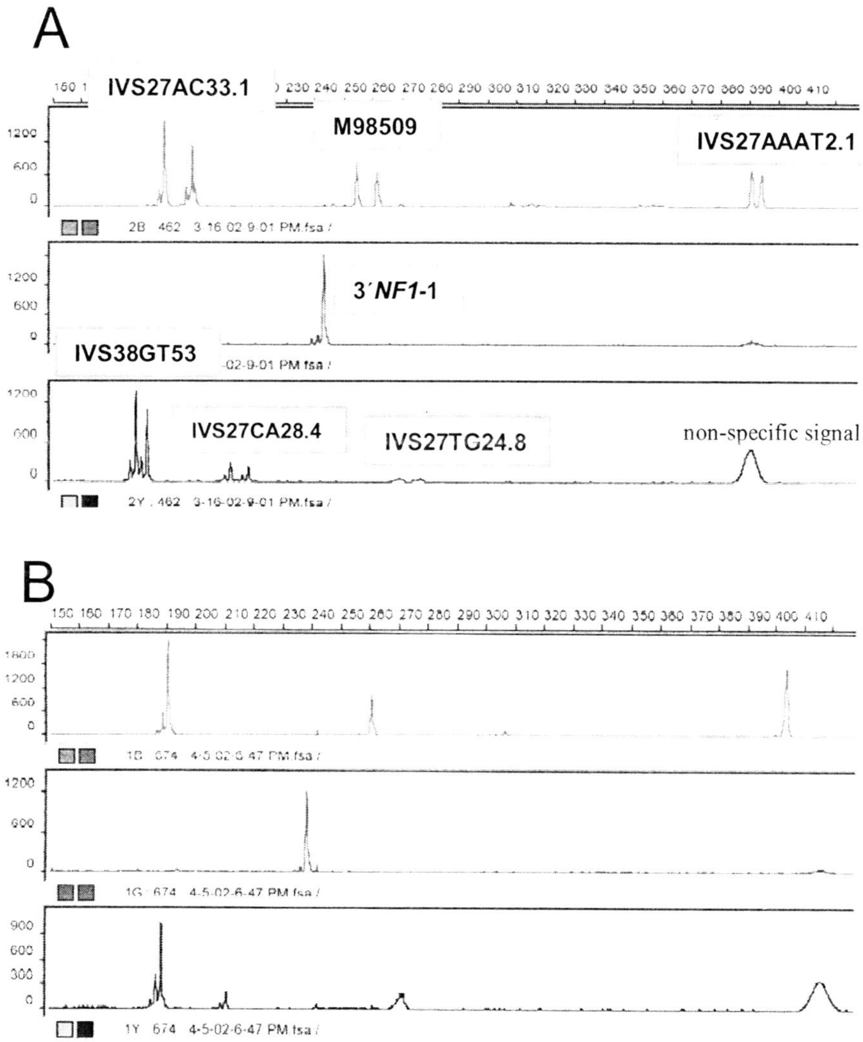

Figure 5. Examples of screening large deletions in the *NF1* gene region using 7 microsatellite markers. **A**: heterozygosity (two peaks) of 6 markers excluded this patient for having deletion covering the region. **B**: all 7 microsatellites were apparently homozygous, indicating high probability of monosomy and thus heterozygous deletion of the region in this patient.

Method

Materials

1. QIAamp DNA Blood Mini Kit (Qiagen)
2. 100 mM dNTPs (Invitrogen)
3. TaqGold polymerase, 10x buffer with 1.5 mM MgCl (Perkin-Elmer)
4. Primers for 7 microsatellite markers for the NF1 gene

5. Hi-Di™ Formamide (Appliedbiosystems, Cat. No. 4311320). This deionized formamide is recommended for sample suspension for electrokinetic injection on capillary electrophoresis systems. Aliquot and store at -20°C.
6. GeneScan-500 [Rox], Cat. No. 401734 (Appliedbiosystems)
7. GeneScan capillaries, 47cm x 50μm, Cat. 402839 (Appliedbiosystems)
8. Performance Optimized Polymer 4 (POP-4), Cat. No. 402838 (Appliedbiosystems)

Protocol

(1) Extract DNA from 200 μl whole blood using the QIAamp DNA Blood Mini Kit following the instruction of the manifecturer. Dilute DNAs 5 fold for PCR and store at -20°C.

(2) Set-up 10 μl PCR reaction to amplify each primer.

- 1 μl 10 x PCR buffer
- 1 μl mixed dNTP (25 mM each)
- 1 μl primer (sense + antisense, 5 pmol/μl)
- 5 μl water
- 0.05μl TaqGold (for hot-start, this enzyme is inactive at the beginning and will be activated after 16 min heating at 94°C)
- 2 μl DNA

Make up master mixture containing all components except DNA and primers. For example, for 50 samples, prepare master mixture for 52 samples containing 52 μl buffer, 52 μl dNTP, 260 μl water and 2.6 μl TagGold. Aliquot 7 μl to each tube or well of a plate, add 2 μl DNA and 1 μl primer solution to each sample. To avoid evaporation, add 12 μl wax or oil to each tube or well. Check each tube/well for possible air bubbles which should be removed by either centrifugation or pipetting. We found electric pipettes with dispensing function very helpful for setting up PCR for large number of samples.

(3) Carry out amplification as following:

- Step 1: 94°C x 16 min (to activate the TagGold polymerase)
- Step 2: 94°C x 30 sec
- Step 3: 55-50°C x 30 sec
- Step 4: 72°C x 1 min (back to step 2, cycle 35 times)
- Step 5: 72°C x 10 min

(4) After amplification, pool 1 μl of each PCR product together, add 20 μl HiDi formamid, 0.5 μl Rox 500. The mixtures are heated at 90°C for 2 min and immediately chilled on ice.

(5) For analysis on a sequencer ABI 310, use filter A and a 47 cm capillary. The standard run time is 24 min which can be reduced if the microsatellites are shorter than 200 bp and has to be increased if they are longer than 400 bp.

2. Study Loss of Heterozygosity (LOH) in Tumors and Further Application of LOH Data

Loss or gain of chromosome material are frequent events in tumors. For tumor suppressor genes, the 2nd or somatic alteration toward tumor development is often loss of heterozygosity (LOH). LOH represents either true loss and thus monosomy or isodisomy of the region. Monosomy is correspond to heterozygous deletion or complete loss of one chromosome while isodisomy is caused by somatic recombination. In either case, one allele, usually the healthy allele is lost in the tumor while the allele bearing the first mutation remains either monosomically or isodisomically.

Literally, "loss of heterozygosity" means that an individual is heterozygous for a locus and this heterozygosity is lost in certain tissue, often tumors. Therefore, pared blood – tumor specimen are required for a LOH analysis for the tumor, and at least one of markers linked to the region of interest has to be heterozygous. Using multiple markers increase the probability that at least one of them is heterozygous. Buccal cells can be used an alternative to blood as constitutional non-tumor tissue. The advantage of buccal cells is that they can be obtained without intervention and can be taken by patients themselves. There are also cases, in which neither blood nor other non-tumor tissues are available. For example, patients with malignant tumors may not be alive any more when their tumors are studied for LOH. In such cases, it is often also possible to mark non-tumor tissues in the tumor section under microscope. For experienced genetist, it is also possible to judge LOH without constitutional tissues based on the trace of the lost alleles. However, such judgement is difficult to understand and thus will often not be accepted by others.

Usually, microsatellite typing for LOH is carried out in paired blood and tumor simultaneously. Using fluorescence labelling, it is also possible to type the non-tumor tissues first. The advantage is that, microsatellites which are homozygous in the non-tumor constitutional tissues can be excluded from further tumor analysis. However, ff radio-isotop labelling and denaturing gel is used, the amplified microsatellites from pared samples have to be loaded side by side on the same gel in order to be able to compare.

Judging LOH very much depends on experience. Though there are various criteria based on calculating the relative intensity of the amplified fragments, we found the most reliable way is still the direct inspection of the electrogram. Fig. 6 shows an example of LOH analysis using 4 microsatellites linked to the *NF2* gene on chromosome 22. As shown in panel D of Fig. 6, LOH is often not complete. This is evident by the trace of the lost alleles of each marker. Incomplete LOH, or partial LOH is due to the presence of non-tumor cells within the tumor. For example, blood vessels embedded in the tumor are not tumor cells and do not carry the genetic alteration, in this case, LOH. Plexiform neurofibromas often infiltrate normal tissue. The removed tumors thus contain large proportion of non-tumor tissues. The amount of non-tumor tissues can be so heavy that the LOH is masked. The best way to avoid this problem is to prepare a series of sections from a tumor and check the first one and the last one histopathologically to make sure that these sections contain reasonable proportion of tumor cells. The sections in between can then be used for DNA extraction. If no pathologist is available, DNA should be extracted from different parts of a tumor. This may increase the probability that one of the preparation has enough amount of tumor cells.

When using DNA of poor quality or quantity, for example, those extracted from achieved specimen, the amounts of two microsatellite alleles amplified can be out of balance, with similar appearance as partial LOH. This can be checked by analyzing several microsatellites from other loci where not LOH is expected. If those microsatellites show apparent partial LOH, then it is due to imbalance of the alleles in the starting DNA, but not true LOH.

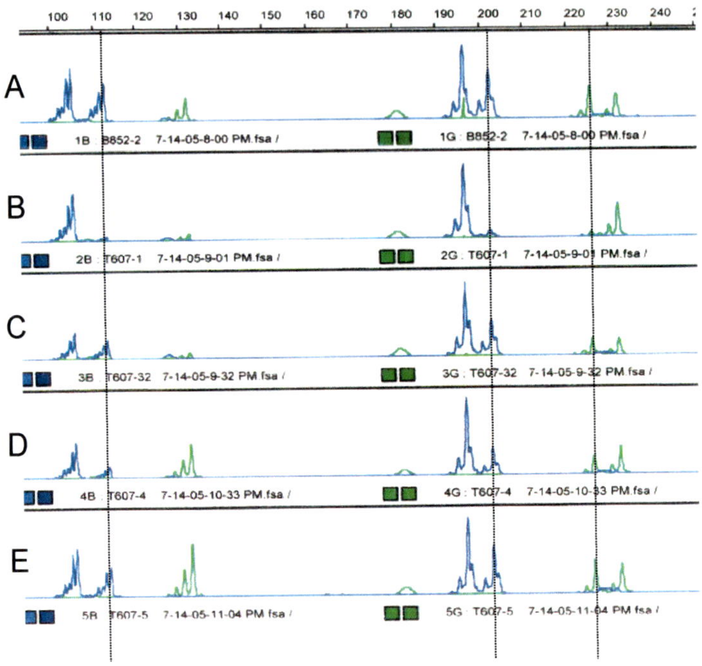

Figure 6. An example of the result of LOH analysis. 4 microsatellites linked to the *NF2* tumor suppressor gene were amplified from DNA of blood (panel A) and 5 tumors (panel B-E) of a NF2 patient. The 1st, 3rd and 4th microsatellite from left each exhibited two peaks. These 3 markers were thus heterozygous and informative in this patient. In contrast, the 2nd marker from the left had only one peak, indicating that both allele of this marker has the same length in this patient. It is thus homozygous and uninformative. A broken line is drawn through the alleles which were lost in the tumors. In the first tumor (panel B), one allele is clearly lost. This allele is presented by the larger allele of the 1st and 3rd markers from the left and the smaller allele of the 4th marker. In the 3rd tumor (panel D), the LOH is not complete, but rather partial. Partial LOH is due to the exist of non-tumor cells in a tumor, for example, blood vessels. Notice that the lost alleles is the same in all tumors with LOH. In the tumors in panel C and E, no clear LOH is visible.

Application of LOH-Data

2.1. Tracing Parental Origin of Mutant Allele in Tumor Suppressor Genes

According to the "two hits" model for tumor suppressor syndromes, the first or constitutional event is small intragenic mutations while the 2nd or somatic events are often LOH of the tumor suppressor gene. Identification the lost allele in tumors thus enables tracing the parental origin of the allele bearing a new germline mutation in a *de novo* patient, provided that genetic materials from at least one of unaffected parents are available. Using this approach, we have determined parental origin of mutation-bearing alleles in 45 *de novo* neurofibromatosis 2 (NF2) patients (Kluwe et al. 2000). Two examples are given in Fig. 7.

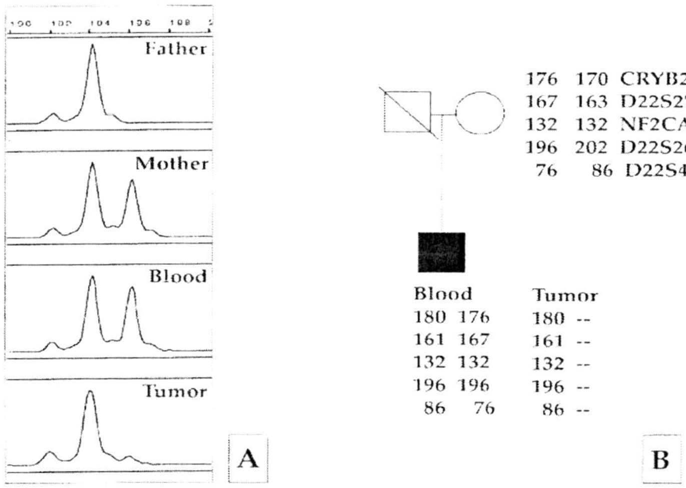

Figure 7. LOH data for tracing parental origin of the mutant allele in NF2. **A**: electrogram of microsatellite D22S268 in blood and tumor specimens from a *de novo* NF2 patient and the unaffected parents. The tumor has lost the maternal allele, inferring a paternal origin of the mutant allele. **B**: haplotype constructed using multiple microsatellites flanking the *NF2* gene in another *de novo* NF2 patient. Even though no paternal blood sample was available, we could still identify the paternal allele as the mutation-bearing one. This is because the maternal allele (176-167-132-196-76) which was inherited by the patients as can be seen in the blood, was lost in the tumor. The remained allele (180-161-132-196-86) which carries the mutation must thus be paternal.

Method

Materials

Same as listed in screening for large deletions plus a QIAamp Tissue Kit (Qiagen) for preparation of DNA from tumor specimen.

Since not every tumor has LOH, tumors should be collected as many as possible to increase the probability of LOH in one or more of them. For preparation of DNA from unaffected parents, other somatic tissues such as buccal cells can be used instead of blood.

Protocol

Extract DNA and amplify microsatellites as described in "screening for large deletions".

2.2. Presymptomatic / Prenatal Diagnosis Using LOH Data

Identification of mutation carriers before clinical symptoms may lead to early detection of subtle abnormalities which may mean better outcome of treatment for the patients. On the other hand, excluding mutation carriers from at-risk individuals means to release them from psychological stress and to spare them from unnecessary examinations which are often very expensive.

The current state of the art for presymtomatic / prenatal diagnosis for a tumor suppressor gene syndrome is mutation analysis of the affected individual. The main drawback is that mutation analysis of the whole gene is very time consuming and expensive. For example, the neurofibromatosis type 1 gene on chromosome 17 (*NF1* gene) has 57 constitutive exons (Kluwe et al. 2003). A complete analysis of this gene using the known mutation analysis takes more than two months. In addition, mutation can not be found in all cases. For example, in neurofibromatosis 2 (NF2), mutations can only be found in about 50% of patients, if DNA from blood is used. The failure of finding mutations in blood of NF2 patients turned out to be due to frequent mosaicism (Kluwe and Mautner 1998).

An alternative way for presymptomatic /prenatal diagnosis is segregation analysis using linked markers. Microsatellites are often used as markers for such analysis. Segregation analysis is based on identification of the disease-bearing allele. To do this, at least two ascertained affected individuals and at least one unaffected individuals from one family are required. Once the mutation-bearing allele is identified, at-risk individuals can be examined if they have this allele. However, large families with multiple affected members are rare. In the case of NF1 and NF2, majority of the patients are *de novo* and thus do not have further affected member in the family. Even in inherited cases, it is often not possible to obtain DNA samples from multiple family members for various reason. Thus, application of segregation is not much more than exceptional.

As described for tracing the parental origin, in tumor suppressor gene syndromes, there is another possibility to identify the mutation-carrying alleles. This can be applied for presymptomatic / prenatal diagnosis. As described for tracing the parental origin of the mutation-bearing allele, identification of the lost tumor suppressor gene alleles enables us to know which of the two alleles carries the germline mutation. One of the two alleles will be passed to the next generation. We can then use the same markers to construct the haplotype in the at-risk individuals who are subjected to presymptomatic / prenatal diagnosis. An example in Fig. 8 also illustrates the principle of this diagnostic method.

This LOH-based method have various advantages over mutation analysis and segregation analysis, for example:

1. It is simple and consists only analysis of several microsatellite marker and can thus be carried out very quickly. Usually such analysis can be done within a week. In urgent cases, for example, prenatal diagnosis, the procedure can also be further accelerated to approximately two days. In comparison, mutation analysis of the *NF1* gene needs more than two month.
2. Only very small amount of DNA are needed, much less than for a mutation analysis. By designing the fragments containing microsatellites shorter than 300 bp, achieved material can also often be used, for example, tumors embedded in paraffin blocks.
3. The LOH-based test is inexpensive. Especially in comparison to mutation analysis of a large gene. It is thus especially attractive for cases in which the insurance does not cover the mutation analysis.
4. A particular advantage is for cases where the mutation can not be found in the affected person.

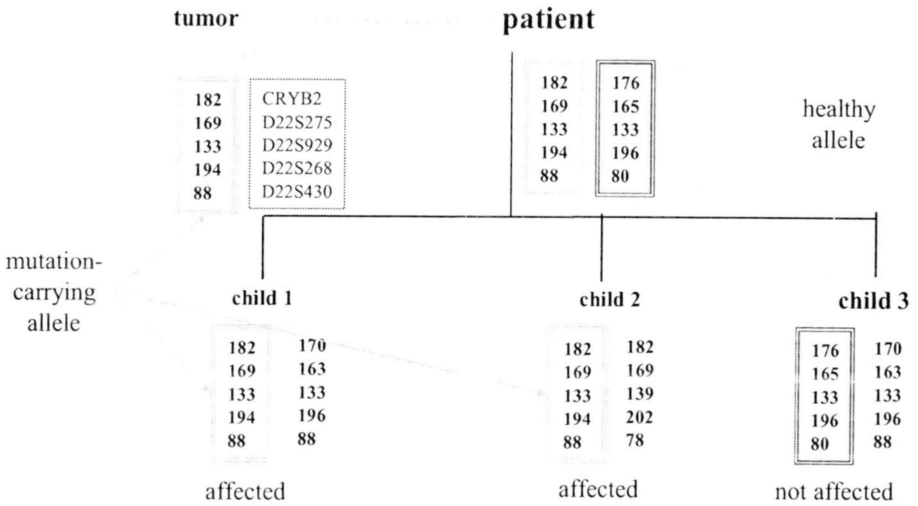

Figure 8. LOH data for determining the mutation-bearing alleles in presymptomatic / prenatal diagnosis. Five microsatellites flanking or within the *NF2* gene were used as markers for alleles. A LOH was found in the tumor of the patient since the allele "176-165-133-196-80" was lost in the tumor. The retained allele "182-169-133-194-88" carries the constitutional mutation in the *NF2* gene. This patient has three children. Two of them inherited the "182-169-133-194-88" from the affected parent, and thus carry the genetic defect. In contrast, the 3rd child inherited the allele "176-165-133-196-80" which was lost in the tumor of the patient. This child thus does not have the genetic defect in the *NF2* gene. The other alleles "170-163-133-196-88" in child 1 and child 3, and "182-169-139-202-78" in child 2 were inherited from the other unaffected parent. Notice that the 3rd marker D22S929 has length of 133 bp in both allele in the patient. This marker was thus uninformative in this case.

5. Unlike segregation analysis, the LOH-based test can be also be applied for *de novo* cases, that is, for patient without further affected family members.

Major disadvantages of this LOH-based diagnosis are (1) the limited availability of tumors and (2) that it can only by applied for cases with LOH in at least one tumor. To overcome these disadvantage, it is important to obtain as many tumors as possible. This increases the probability of finding LOH in one of the tumors. In the case of NF1, skin neurofibromas are present in most patients and can be obtained easily by simple external surgery. Special caution should be taken to avoid wrong interpretation of imbalanced alleles due to poor DNA quality or quantity. Since no LOH is expected in microsatellites in unrelated regions, they can be used as a control.

2.3. Determining Cellular Origin of Tumor Cells

Tumors often consists of more than one type of cells. For example, neurofibromas from NF1 patients contain Schwann cells, fibroblasts, mast cells, perineurial cells and neurons. Schwann cells have been considered to be the progenitors of the tumors. However, there were also data showing involvement of fibroblasts. Direct evidence of the cellular origin for this kind of tumors has been missing.

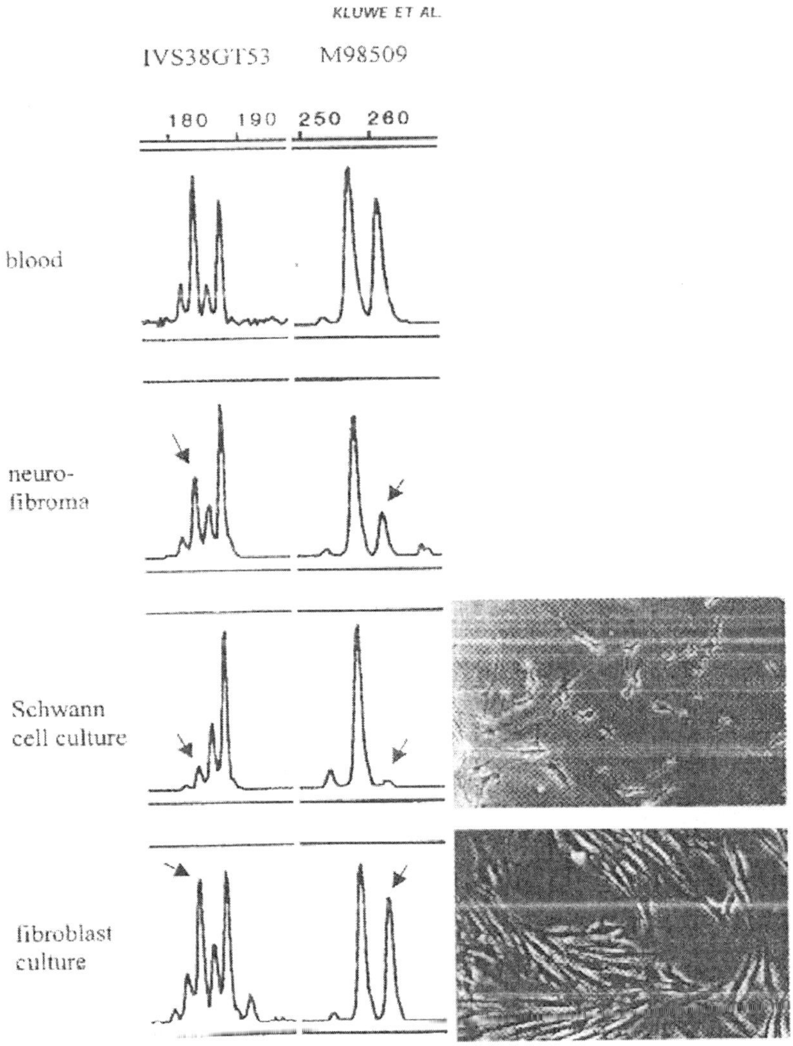

Figure 9. LOH for determining the cellular origin in a tumor (neurofibroma). An incomplete LOH was found in the tumor. In derived Schwann cells, this LOH was even more prominent. In contrast, no LOH was seen in the derived fibroblasts. These data demonstrate that the genetic alteration occurred only in Schwann cells which is a solid evidence that these cells are the progenitor of the tumor. The lost alleles are indicated by arrows. Schwann cells and fibroblasts exhibited different morphology (insert).

When a neurofibroma is dissected and cultured under standard condition, fibroblasts grow predominantly, further suggesting that the role of these cells in tumor growth. Under special cultural condition, it is also possible in some cases to selectively grow Schwann cells from neurofibromas. We dissected a neurofibroma enzymatically and divided the suspension containing cells into two parts. One was cultured under standard condition, the other under condition optimized for Schwann cells. As shown in the inserts in Fig. 9, cells obtained under condition for Schwann cells exhibited morphology of Schwann cells: small, bi-poplar and often shiny cell bodies. In contrast, cells grown under standard condition exhibited typical features of fibroblasts: flat and thin. DNA was extracted from these cells separately and subjected to microsatellite analysis. The results are shown in Fig. 9. Two microsatellite markers within the *NF1* gene were informative for this patient, as can be seen as two peaks/marker in the blood DNA. Loss of heterozygosity of both the markers was found in the tumor, though the loss was incomplete. The same LOH was found in the enriched Schwann cells, and the loss was complete indicating that this culture contains pure tumor cells. In contrast, the LOH was not found in DNA extracted from the fibroblasts grown under standard cultural condition. This result was the first solid genetic evidence that only the Schwann cells carry the somatic alteration in the *NF1* gene. These are thus the progenitor tumor cells in neurofibromas. In contrast, fibroblasts are not directly involved in the tumorigenesis, though they may play some promoting role (Kluwe et al. 1999).

Our study also demonstrated that tracing LOH is a simple and efficient way to examine if cells obtained in culture really are tumor cells which must carry the same genetic alteration as the tumors.

2.4. Verifying Tumor Cells Obtained in Culture

When a tumor is dissected and some cells grown in vitro, one would expect that these cells are tumor cells. However, this is often not the case. As demonstrated in our above study, fibroblasts grow faster than Schwann cells from a neurofibroma, though the Schwann cells are the tumor cells and fibroblasts not. Even for malignant peripheral nerve sheath tumors (MPNST), a very aggressive cancer, we have demonstrated that cells grown in culture are not always the tumor cells (Frahm et al. 2004).

To examine the identity of cells in culture, we can use the LOH finding in the tumors. For example, LOH of tumor suppressor genes *p53* and *p16* can frequently be found in many types of malignant tumors. Microsatellites linked to these genes can be used as markers (table 3).

DNA need to be prepared from the original tumor, cultured cells and some non-tumor tissue of the corresponding patient. Next, microsatellite markers are amplified from these DNA. To compare the markers in blood, tumor and cells, corresponding electrograms are arranged on top of each other, as the example in Fig. 10. In this example, two microsatellite markers are shown, one linked to the *p53* gene and the other for the *NF1* gene. Both tumors (MPNST1 and MPNST2) showed LOH for the two markers. The same LOH was also visible in cells derived from MPNST 1 on the left. This ascertained that the cells carry the same somatic alteration as their original tumor. They are thus the tumor cells. In contrast, cells derived from MPNST 2 (on the right) did not have the LOH which was present in the original tumor. These cells do not share the genetic alteration with the tumor. Thus they do not represent the tumor cells.

This LOH-based verification is only possible if LOH of any relevant gene can be found in the original tumor.

Table 3. Microsatellite markers linked to *p53* and *p16*

Linked to	Marker	Size range	Labelling	Color (FilterA)	Location	Primer Seq
P53	**D17S520**	130-144	JOE	Green	17p	GGAGAAAGTGATACAAGGGA TAGTTAGATTAATACCCACC
P53	D17S796	144-174	6-FAM	Blue	17p	CAATGGAACCAAATGTGGTC AGTCCGATAATGCCAGGATG
P53	D17S804	156-170	JOE	Green	17p	GCCTGTGCTGCTGATAACC CACTGTGATGAGATGTCATTCC
P53	P53invl	105-140	Tamra	Black	17p	GCACTTTCCTCAACTCTACA AACAGCTCCTTTAATGGCAG
P16	D9S168	227-247	Tamra	Black	9p	GGTTTGTGGTCTTTGTAAGG TGGTTTGTTTGTATAACTATCATTG
P16	D9S171	157-179	6-FAM	Blue	9p	AGCTAAGTGAACCTCATCTCTGTCT ACCCTAGCACTGATGGTATAGTCT
P16	D9S157	120-135	6-FAM	Blue	9p	AGCAAGGCAAGCCACATTTC TGGGGATGCCCAGATAACTATATC

The annealing temperature was 55°C for all.

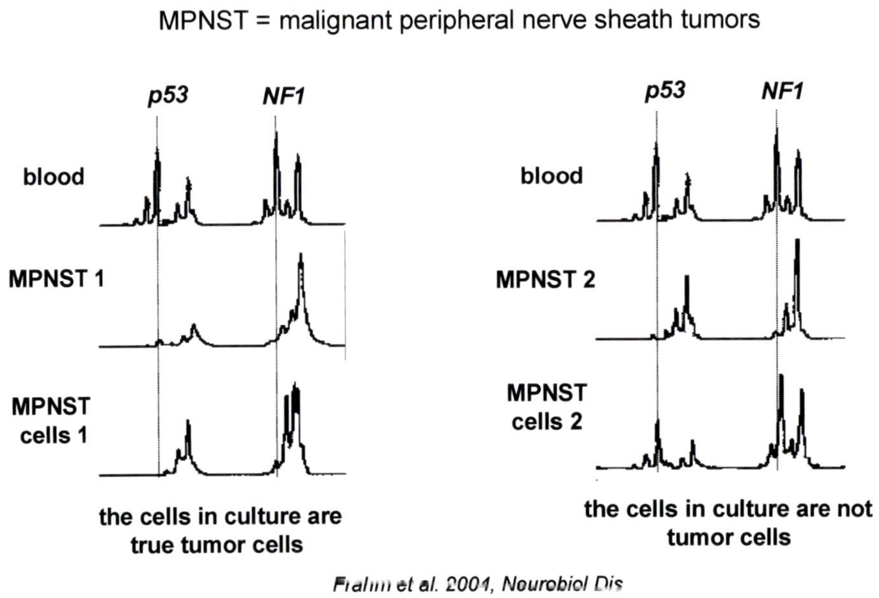

Figure 10. LOH of one *p53* and one *NF1* markers was found both tumors MPNST-1 and MPNST-2. Cells derived from MPNST-1 exhibited the same LOH, indicating that the obtained cells are tumor cells. In contrast, no LOH was seen in cells derived from the MPNST-2, demonstrating that they are not neoplastic. One possibility is that fibroblasts grown faster than the tumor cells.

Conclusion

This chapter introduced several extended application of using microsatellites as markers. These examples demonstrated that microsatellites are easy to use and are often very effective for screening large deletions and characterize cells in culture. Despite the rapid development of numerous assays for SNPs, microsatellites will remain a powerful tool for genetic studies in the future.

References

Frahm, S; Mautner, V.-F; Brems, H; Legius, E; Debiec-Rychter, M; Friedrich, RE; Knöfl, W; Peiper, M; Kluwe, L. Genetic and phenotypic characterization of tumor cells derived from malignant peripheral nerve sheath tumors of neurofibromatosis type 1 patients. *Neurobiol Dis* 2004, 16, 85-91

Kluwe, L; Mautner, VF. Mosaicism in sporadic neurofibromatosis 2 patients. *Hum Mol Genet* 1998, 7, 2051-2055

Kluwe, L; Friedrich, R; Mautner, V-F. Loss of NF1 alleles in Schwann cells but not in fibroblasts derived from an NF1-associated neurofibroma. Gene, *Chrosmosom and Cancer* 1999, 24, 283-285

Kluwe, L; Mautner, VF; Parry, DM; Jacoby, LB; Baser, M; Gusella, J; Davis, K; Stavrou, D; MacCollin, M. Parental origin of de novo mutations in sporadic neurofibromatosis 2 patients. *Neurogenet* 2000, 3, 17-24

Kluwe, L; Siebert, R; Gesk, S; Friedrich, RE; Tinschert, S; Kehrer-Sawatzki, H; Mautner, V-F. Screening 500 unselected neurofibromatosis 1 patients for deletions of the NF1 gene. *Hum Mut* 2004, 23, 111-116

Krawczak, M; Cooper, DN. The human gene mutation database. *Trends Genet* 1997, 13, 121-122

Index

D

E

F

G

H

I

J

K

L

M

N

O

S

T

U

V

W

Y